# THE JOURNEY OF MIND

# THE JOURNEY OF MIND

*Evolution of Knowledge Consciousness*

## RAJESH JAYARAM

PARTRIDGE
A Penguin Random House Company

Copyright © 2013 by Rajesh Jayaram.

ISBN:       Hardcover          978-1-4828-1372-2
              Softcover          978-1-4828-1370-8
              Ebook              978-1-4828-1371-5

**To order additional copies of this book, contact**
Partridge India
000 800 10062 62
www.partridgepublishing.com/india
orders.india@partridgepublishing.com

# ROUTE MAP

# PRAISE FOR "THE JOURNEY OF MIND"

*"Author's obvious notion and insights made me realise the mistakes i was exactly doing without even cared to think, after reading this book I rearranged financial portfolios to safeguard my financial investments"*
                                   *—Shiva subramaniam, Shares Trader and financial Broker*

*"Excellent information about MANUVANTARA the Indian cosmic time revealed finally and names of forthcoming incarnations as per hindu religion were scripted in Telescope thinkers . . . ."*
                                   *—Robert cliffe, Historian and Anthropologist.*

*"Hinduism and Metaphysics comparative explanation specifically in cosmological science is fabulous reveals the deep understanding and the determination displayed in narration and is appealing . . . ."*
                                   *—Raghavan sundaramurthy, Professor in Quantum Physics*

*"Young authors from india emerging and reaching global audience, gaining larger significance towards indian religion, art, culture, philosophical ideas and revealing the significance and popularity of monumental heritage and its values"*
                                   *—Samantha Arthur, Coloumnist and Blogger*

*"masterpiece work . . . . the informations about ancient wisdom and modern science is in correct proportion and categorically explained nothing short of the genious work"*
                                   *—Gakushi Sadahiro, ex-president, Japanese chamber of cultural science.*

*". . . . description of purana's part for younger generation to understand is rightly interpreted in scientific view with a common sense"*
                                   *—Shivlal pandey, Hindu activist*

*"Journey of mind the second chapter is an eye opener, it made me realize that iam living in a very small world and decided to take this message worldwide"*
                                   *—Clifford James, Consultant, Publishing services*

My Dear,

I always gifted you best books available every year as Christmas gift, I knew very well that you always had curious mind in obtaining intricate details, I enjoyed reading this book which changed my perception and thoughts, which gave me, a new way of approaching and dealing matters both in personal and professional life, as I immediately remembered you of gifting this book to you, I am sure you will enjoy it and share this knowledge with your friends and someday to your children whole heartedly . . . & New Year greetings . . . .

—With love . . .

# IN THE WORDS OF AUTHOR

I, Salute to the Indian Idealogists, Rishis, Yogis and Modern Scholars of thousands of years, who has been continuously been the centric magnets of retelling the ancient Indian History, Religion and Civilization to Indian people, made Hinduism one of the most oldest, powerful and *Scientific* Religions in the world sustaining and developing atleast 2.165 million years from today, following the VEDAS (as I consider the VEDAS are not subject to the dating validations, as the VEDAS is the Sacred Divine Hymns SHRUTI and SMRITI Samhitas has been passed to generations linguistically, before scripting in SANSKRIT Language) Scholars & Authors registering literary to their credits dating back from Ancient India starting from Brahmarishis, Valmiki, Veda Vyasa to modern day A.C. Bhaktivedanta, Swami Prabhupada.

The attempt of this book is to present PRECISE AND CONCISE accounts of Premedival era and Medival era of world civilisations being influenced by Ancient Indian Idealogies and Vedas shall give the interesting perspective to the readers of Modern Science, the scripted Vedic Samhitas lending the theoretical base to modern science discoveries in the field of Aero Dynamics, Medicine, Particle physics etc., Western Scientists like Einstein, Newton and modern Nobel laureates in Physics have accepted that the Vedas were Foreword and Sparks of the idea which helped their Inventions and Discoveries, remembering Einstein in his book mentioning "that we owe a lot to the Indians who taught us how to count, without which no worthwhile scientific discovery could have been made", However some treated Vedas as Primitive rituals, The 18th & 19th Century Professors Max Mueller, TH.Griffith, HH Wilson, William Whitney and Voltaire realised that the gap is bridged by referring Vedas being the mightiest literary works of the Human History.

My Objective to write this book is to enlighten, the younger generations today reaching boundaries of the continents in various fields of Science, Engineering and Technology should be aware of these foundations and rooting to ANCIENT INDIAN VEDAS(Facts) which taught the world about the subject of Creation of Universe (Metaphysics)-RIG VEDA, Principles, Art, Culture, Society, (SAMAVEDA), Leadership, Social Structure & Practices, Community, Arithmetics and Ritual Science (YAJUR VEDA) and further Public Administration, Weaponry, Chemicals and Psychology and Economics in (ATHARVANA VEDA), Atharvana Veda being the Last and final part of VEDA which explains the facts earlier missed and explains more in detail to common man understanding level which scripted in earlier RIG, SAMA and YAJUR VEDAS.

This work is dedicated to beloved mother Late Shri. Saraswathi Ranganathan who taught me the mythology and science dwelling together to dominate this world and human society millennium after millennium till date, Thankful to my loving brothers and other family members who has been supportive all along in collecting Research materials, Reference Manuals, and Friends from Libraries providing most valuable Prints.

# EVOLUTION OF KNOWLEDGE CONSCIOUSNESS

## into the concept of

# ANCIENT WISDOM AND MODERN SCIENCE

# 1

# *As the Journey of Mind begins . . . .*

I t all started on a beautiful winter evening in december last year while sitting at my balcony alone under canopy, as i was tasting the hot brazilian crushed coffee bean and drizzles of sugar in milk, just lifted the wonderful aroma, made me close my eyes, and when I opened the tranquile beauty of ootacamund hill view with a fantastic view across lush green grasslands, did glittered adding to its beauty, while sunset on the west among the clouds behind the hill, with a cold breeze gave thoughts what in life can anybody could miss something like this, as doorbell rang I remembered about the visit of my friends that evening and the dense fog carried away our troop to discussions and diverted to various topics, after a couple of hours we had dinner and one my friend reminded me "hey rajesh, night is getting colder man? and i need to leave, man i had great evening today, he friendly insisted the group we should get along like once a week atleast, we should make it a strict rule that we friends will join every Sunday between 4 and 6, and the funniest in our group agreed to it telling "yeah it's a wolfgang promise" . . . . and we all took part.

<<<<<<< §µΨΦ∞ЖλπЦδ >>>>>>>

Something flashed my mind upon taking my face towel about the book which I bought yesterday about the rare medicinal plants and the extinction of its species, and I took myself on the bed with that book slowly sneaking about the book and the author's biography, which made me impress the education background and his field experience, after reading few chapters, I realized something as this author expressing reprimand towards the growing era of casual attitude among people today, I honestly felt inside me, it cannot be avoided that's how it happens as that's the chain connection which will continue towards the era of knowledge begins,

Something alarmed me really at that moment, if in case the upcoming era of knowledge is well provided with the "macro knowledge in micro categories", and the next generation's responsibility level, is necessarily to be triggered to avoid this growing trend of casual attitude which is spoiling the nation into corruption is the leading no.1 evil which many governments in the world is elusive to tackle, if not addressesd at right time, may take an another chain reaction, casual attitude –> corruption –> greedy minds –> you don't want to know what's next . . . .

I decided to take on a journey of writing a book which i realized that it should communicate to anybody who can understand it, but it took the micro categories of society in the path of man's approach to life. In technical terms is called as "Anthropology", study about humans.

<<<<<<< §µΨΦ∞ӜλπႱδ >>>>>>>

**Preparing for the journey** . . . .

I carefully prepared a check list of essentials which is required for the wonderful journey and I took the name cards of renowned think tanks, historians, archeologists, geologists, professors, librarians, I double checked to make sure that i have all the friends contact numbers in mobile phones, and iam determined ever before and decided on making this journey, as i knew very well as holiday season nearing from last week of December, I should plan the journey accordingly that I should not miss my schedules on meeting people, perhaps i could convince some of my friends to spend a couple of days to accompany me to few places around, if required,

<<<<<<< §µΨΦ∞ӜλπႱδ >>>>>>>

This objective of this journey is not to give readers a fiction or fantasy, it is all about scientific, hidden, rare, significant, researchful, historical, philosophical, idealogical and derived informations. This book is for curious minds continuously thriving to become a learned wise man. This had been a fantastic journey to me and the people who are the part of making this journey a success have enjoyed it with me,

*portraying the experience of the journey to others* . . . .

**I decided to have a telescope(narration style) to this story, to the viewers(readers) for the first time to visualize the nonfiction book, however i have decided to adopt contemplated approach of writing to break all the formulised version of writing a book.**

The days when i was collecting informations and datas for this book and potentials for validations indicated me i should be double sure that the Ancient wisdom have ruled this face of earth for millenniums and it even continued in asian countries till date,

Many people all around the world, wondered the asians knew something more which we did not really knew and what they could not understand, thats why asians worshipped the Lord almighty in human form and respected saints unlike like Islam (book of Quaran) and Christianity (image of Cross) were the religion of scholar, the religion followed the commandments of one messenger, and westerners always harangue at Indians particulary looking at a religion of so many gods,

India is the land of proud owner of perpetual Ancient Vedic Manuscripts, the so called Rishis and Yogis contributed to the religion comprehending it the most scientific, the wisdom derived from the Vedas were pure, unchangeable, undeniable, accepted scientific facts, extended over millions of years atleast passing the lore generation after generation.

About Idol worship, Now a reader should think how a scientific religion ended up in worshipping the idol or form? which again falls into category of "messenger's commandments followed religion"?

**The primitive people had a form of trust of the formless Lord almighty not visible to the human eye**, Rishis and Yogis used an oblation of using this form of trust oblating humans with a human, the idea of god being a human . . . . which resulted in primitive men in repugnance, permeates, lucid and luminous nature in understanding the concept of god and resulted in equanimity and great restraint.

<<<<<<< §µΨΦ∞Җλπцδ >>>>>>>

**The attuned and emancipated approach in the idea of Divinity**:

The difference in approach or naming the god, Islamic and Christianity the major religions acknowledged their leader as messenger, Indians in Hinduism had it restrained the same concept as incarnation of lord himself, who was born over a period of time when required on earth, for world's well being and the righteousness.

Hinduism also attuned the messengers separately, Hinduism all over history recognized and offered them the highest position as saints who liberated themselves from being in married life and other worldly pleasures and took the pains of personal sacrifices for the divine belief and teaching people about divinity (Vedanta, Commentaries on actual 4 book of Vedas-scientific facts) and Hinduism was never ashamed of putting their biography in records as it is without hiding anything . . . .

Never Hinduism did the mistake of promoting the superfluous of religious leaders or messengers, it only cherished and admired the gurus, scholars and messengers who practiced the sanyasa, and became saints who had liberated themselves of this worldly rules and to them, Indian people given them a colomn in the history for the inequable services they did for proposing the view points of vedas and vedantas, ultimately taught the Vedas.

Vedas only taught the philosophy and idealized the ancient science into hindu temples and in hindu puranas (epic stories) and the Upanishads (Commentaries or discourse of epics, Scientific and philosophical ideas), influencing most of the asian countries in terms of divine belief and worship practices, art, culture, dress, respect and behavior system, the approach of life, living as a society and community of assimilation.

<<<<<<< §μΨΦ∞Ж入πЦδ >>>>>>>

**Purpose of this topic into this book's journey**:

For the younger generation and over the concern of growing nature of casual attitude in people's mind and behavior should be eradicated or atleast to be reduced to significant level, only possibility is by providing the potential and useful knowledge to the readers like you with curious mind willing to enlighten the casual goodies, fundamental baddies, courageous idiots, angry fast mouth runners, stealers, heart fillers of jealous and innocent future generations and i don't intend to mention murderers in my list.

I welcome you all to have panaromic view thru my telescope where you can vision the twilight zone of knowledge, wisdom, truth and fact . . . .

<<<<<<< §μΨΦ∞Ж入πЦδ >>>>>>>

# 2

# *Ancient India and its Idealogies*

L ets begin with an understandable note before we travel into the Journey of mind and explore its thinking . . . . and on proceeding we will try and understand what western scientists in their own words told about Indian Vedic philosophies.

*Albert Einstein (1879-1955), Physicist, Nobel Laureate: When I read the Bhagavad-Gita and reflect about how God created this universe everything else seems so superfluous.*

*W. Heisenberg (1901-1976), German Physicist and Nobel Laureate: After the conversations about Indian philosophy, some of the ideas of Quantum Physics that had seemed so crazy suddenly made much more sense.*

**Dick Teresi, American author of 'Lost Discoveries':** . . . . *Some <u>one thousand years before Aristotle</u>, the Vedic Aryans asserted that the earth was round and circled the sun . . . . <u>Two thousand years before Pythagoras</u>, philosophers in northern India had understood that gravitation held the solar system together, and that therefore the sun, the most massive object, had to be at its center . . . . <u>Twenty-four centuries before Isaac Newton</u>, the Hindu Rig-Veda asserted that gravitation held the universe together . . . . **<u>The Sanskrit speaking Aryans subscribed to the idea of a spherical earth in an era when the Greeks believed in a flat one . . . .</u>** The <u>Indians of the fifth century A.D. calculated the age of the earth as 4.3 billion years; scientists in 19th century England were convinced it was 100 million years . . .</u>*

**Philip Goldberg**: *The best evidence of this is science's response to the religions of the East over the course of the last 200 years.* **As the French Nobel laureate Romain Rolland said early in the 20th century, "Religious faith in the case of the Hindus has never been allowed to run counter to scientific laws." The same can be said for Buddhism, which derives from the same Vedic roots.**

As early as the 1890s, Swami Vivekananda spent time with scientific luminaries such as Lord Kelvin, Hermann von Helmholtz, and Nikola Tesla. "Mr. Tesla thinks he can demonstrate mathematically that force and matter are reducible to potential energy," the swami wrote in a letter to a friend. "I am working a good deal now upon the cosmology and eschatology of Vedanta.

**Hindu and Buddhist descriptions of higher stages of consciousness have expanded psychology's understanding of human development and inspired the formation of provocative new theories of consciousness itself.** Their ancient philosophies have also influenced physicists, among them Erwin Schrödinger, Werner Heisenberg and J. Robert Oppenheimer, who read from the Bhagavad Gita at a memorial service for President Franklin D. Roosevelt.

**Carl Sagan** *called Hinduism the only religion whose time-scale for the universe matches the billions of years documented by modern science. Sagan filmed that segment in a Hindu temple featuring a statue of the Lord Nataraj as the cosmic dancer, an image that now stands in the plaza of the European Organization for Nuclear Research CERN in Geneva.*

<<<<<<< §µΨΦ∞Ж入πЏδ >>>>>>>

What did India had in its book of Vedas to its significance which many of them today in the materialistic world are not aware, but right from the beginning, the moment Vedas are proposed to common man, the continous evolution of knowledge consciousness has happened in the way of life and in the thinking process of humans which led to the understanding of the theories and resulting in phenomenal inventions and discoveries which has been accepted in most parts of the world.

Before dealing with the core concept of comparing and explaining the Indian Vedas (Ancient science) and modern science, it is better to give a small introduction about Indian puranas and upanishads and the contents described in it for explaining others about God and his will, the purpose and the meaning of divine worship. This concept will be supported with a small rundown for few chapters on ancient civilization, belief, culture, influence, practices, trade systems, kingdoms and rulers (Anthropology) and metaphysical concept of cosmic time or space time or divine time with earth time and it has been measured **not in light years but in human years** and after initial understanding about anthropology and assimilations the actual hymns from Vedas and the translated meaning for vedic hymns explained in finer details . . . .

When we try to understand what is commanding any person's drive is what he is believing in, what shall be the basic belief? the concept of God?!, Yes?, but no religion is higher than its concept of God, the level of its height in concept is making the religion stronger, sustaining and developing over the centuries. The change in migration in the past 6 decades across the world is very high in the last 2 centuries, when people started migrating to other countries and people lived under its borders, Nationality, policies, is connecting to the internetwork aspects. A Christian living in a muslim country, a Muslim living in a Hindu Country, connected the various religion thus making this a assimilated world,

World is A GLOBAL VILLAGE now connected with Internet popularly with the name of Facebook, twitter and linked-in and many other web-portals making friends without physical presence.

Sharing experience is what many social networking websites propounds, youngsters from any religion were busy finding their soul mates, jobs, education, influence and everything else, which is giving them enormous success in their life when it comes to their love, career and wealth and I wonder why the divorces happening more to this segment of internet driven community, what and where is something is going wrong, is finally leading to the lack of knowledge with the history and background of the culture, religion what made them and why these individuals in the age of data is not trying to know about this aspect. Okay less maturity, diversions fine, but what makes them stop reading about history?.

Parents were failing to propose the youngsters with CORRECT KNOWLEDGE of their religion and the concept of GOD, Religion is the way of LIFE, so does for every religion, The recent changes and number of divorces going up in the land of INDIA, from where Vedas was taught to the world, the most erotic religion on earth now and then, But what happened??? Money will buy you the meterials and even sex, but will it give you the will of affection??

<<<<<<< §μΨΦ∞Жλπμδ >>>>>>>

In this and in the proceeding Chapters we are going to deal with the lack of Knowledge on following aspects,

What we missed from History?
What we forgot from Civilisation?
Where we misconceived the concept of religion?

What misrepresentations happened and done by who?

What we are failing to pass on and the mistakes we are doing while passing on the knowledge?

The daily changes in life is continuously pushing the man to compromises and to run for survival. In the midst of changing Political situations, Business, Valuations, Relationships, Climatic conditions something is holding up with great intensity is driving the human force is HOPE and TRUST. In other words the concept of God, Divinity etc.,

All along maths and science worked out together in framing the concept or the idea of God got well shaped up to the religions and its teachings. It is everybody's responsibility of knowing what is the concept of god, Hinduism as religion and its properties involved in its idealogy. Ancient India gave us the fantastic scriptures in fact gave the world the way of life and taught how to understand the substance matter of the life thru its Vedas is what we forgot and missed from history?

<<<<<<< §µΨΦ∞Жλπμδ >>>>>>>

In India, everything starts and ends with religion, as we witnessed the moment a person gets his dream fulfilled in his life immediately performs a pooja to his favourite deity, it matters even buying a bike, car, house, a baby born in a family right from marking the birth time and date, to pronouncing a name, to getting admitted to the school to till a person dies the unshakable system of worship and paying respects to the deity rolls on to the day to day life, to Hindus being religious and auspicious matters a lot. The very same behavior and attitude has been seen everywhere in asian countries is found common. While analyzing it deeper it is understood the influence from india is shadowing in most of the asian countries.

We have seen 90% of the homes in india have this "sri yantra" picture in frames and even in work place, business houses, which is certainly shows the proof what the religion and gods meant to Indians.

I have spoken to many westerners about the people of India, they all told in common is india is the place of strong religious land and religion and the auspicious feeling is there in every blood drop in most indians, But what makes Indians so religious, its because the scientific knowledge that Vedas contributes to the religion, to begin with the birth of Vedas (ancient science),

## BIRTH of INDIAN VEDAS:

Perpetual VEDAS (Unchanged Truths/Facts/Sciences)—PROPOUNDED thru BRAHMARISHIS were considered for providing us with Vedas to lead a way of balanced life. How was the Vedas born? As Indian Puranas explains BRAHMA (the creator god) meditated to supreme deity and Lord NARAYANA(the organizer and preserver god) appeared before him and enquired about his sadness clouding his mind, BRAHMA did told Lord NARAYANA that he is creating the universes inside galaxy but without any purpose and objective and enquired Lord NARAYANA why he has been deputed as creator and do creation of universe, puzzled BRAHMA requesting for the need of knowledge and persipicacious, which will fulfill his purpose and objective and Vedas manifested surrounded to BRAHMA's surprise in millions of stars. once again confused BRAHMA requested he doesn't know to what to select upon and astute to Lord NARAYANA almighty, Almighty handpicked 3 Vedas and gave it to BRAHMA, **upon which BRAHMA created the universe and taught the essence and elixir of pranava mantra AUM (the root for existence given by existed).**

Towards the end of each KALPA (relatively long period of time (by human calculation) is COSMIC TIME was normally defined for days in the life of BRAHMA. The life time of BRAHMA or the distance of BRAHMA loka or BRAHMA world to Bhoolaka (EARTH). On observing the COSMIC TIME mentioned in vedas is closely connecting to Modern science Galactical Collision approximately dated, but not in VEDAS, VEDAS were very accurate and even named each and every single day in the life of BRAHMA.

we can also look it as 100 God Years = Life Span of God

The Vishnu Purana scripts the cosmic time or divine time measurement follows: author explains this cosmic time for explaining the distance with human years rather explaining it in solar light years, for the benefit of readers understanding:

4,000 + 400 + 400 = 4,800 divine years (= 1,728,000 human years) = 1 Sat Yuga
3,000 + 300 + 300 = 3,600 divine years (= 1,296,000 human years) = 1 Treta Yuga
2,000 + 200 + 200 = 2,400 divine years (= 864,000 human years) = 1 Dvapara Yuga
1,000 + 100 + 100 = 1,200 divine years (= 432,000 human years) = 1 Kali Yuga
12,000 divine year = 4 Yugas (= 4,320,000 human years) = 1 Maha Yuga
24,000 divine years is 1 revolution of sun around its orbit along with planet family rotating around sun.

Vedas is the first authenticated proposed and antiquated but reliable science in the world, which scripted the rotation and revolution of the sun in its orbit to complete one full circle and its time is measured accurately, Modern science believes that sun is not revolving in any orbit, but the planets of this solar family is rotating and revolving in their orbit, but VEDAS say SUN too rotating and revolving in orbit along with its planets revolving the sun around.

**Reckoning of time for Brahma and conversion of cosmic or space time in human years**

1000 Maha-Yugas = 1 Kalpa = 1 day (day only) of Brahma (2 Kalpas is a day and night of Brahma, 8.64 billion human years)
30 days of Brahma = 1 month of Brahma (259.2 billion human years)
12 months of Brahma = 1 year of Brahma (3.1104 trillion human years)
50 years of Brahma = 1 Parardha
2 parardhas = 100 years of Brahma = 1 Para = 1 Maha-Kalpa (the lifespan of Brahma) (311.04 trillion years)

**Would you like to know the age of Current Brahma?** or the the distance to the earth calculated from Brahmaloka? **to be remembered there are many brahmas will be designated during the life span of Lord Narayana or Lord Vishnu.**

**155.52 Trillion Years** (50 years of Current Brahma) as of 2013 AD in Hindu Cosmic calendar system.

**The current Kali Yuga began at midnight by 18 February in 3102 BC** as per Julian calendar only 5,115 years are passed out of 432,000 years of current Kali Yuga, and hence another 426,985 years are left to complete this 28th Kali Yuga of Vaivaswatha Manvantara. What is Manvantra? we will deal with manvantra and the forthcoming yugas its names and incarnations yet to happen later and it is explained eloborately in the procceding chapters when we deal with deeper cosmic time periods.

As the calculation found today in panchanga (Hindu Daily and time Calendar system) is found to be described in Vedas before thousands of years (when many civilizations did not had their literary language), claiming to be accurate in dating of cosmic events occurring in space, the solar eclipse, lunar eclipse and the starting time and ending time of new moon day and full moon day scripted so accurate is still happening as per calculations of panchanga, One to remember these panchangas were scripted in the days when no telescope was there to see the events occurring in space.

### VEDAS Propounded thru Brahma-rishi

| BRAHMA-RISHIs NAME | Numbers of Vedic R'Cas and Hymns erudited |
|---|---|
| ANGIRAS | 3619 |
| KANVA | 1315 |
| VASISHTA | 1276 |
| VISHWAMITRA | 983 |
| ATRI | 885 |
| BHRGU | 473 |
| KASHYAPA | 415 |
| GRTSAMADA | 401 |
| AGASTYA | 316 |
| BHARATA | 170 |

Seven brahmarishis have the special status of "Saptarishi". They included Kratu, Pulaha, Pulatsya, Atri, Angiras, Vasishta, Bhrigu. In Hindu astronomy, the stars of the Big Dipper are named for the Saptarishi.

INTERSTINGLY THE STATE FLAG OF ALASKA representing the Big Dipper the symbol assigned to the flag lately in May 2, 1927, I was wondering why alaska have chosen the BIG DIPPER Constellation not any other Constellation instead.

| SAPTARISHI Constelallation as per vedas | ALASKA Flag representing BIG DIPPER |
|---|---|
| | |

| SAPTARISHI CONSTELALATION (INDIAN NAME) | BIG DIPPER or URSA MAJOR (WESTERN NAME) |
|---|---|
| KRATU | DUBHE |
| PULAHA | MERAK |
| PULATSYA | PHECDA |
| ATRI | MEGREZ |
| ANGIRAS | ALIOTH |
| VASISHTA | MIZAR |
| BHRIGU | ALKAID |

<<<<<<< §µΨΦ∞Жλπцδ >>>>>>>

## UPAVEDAS (The Branches of Major 4 VEDAS—RIG, YAJUR, SAMA and ATHAVAVEDA)

(AyurVeda—Medicinal Science asssociated to RIG VEDA, explains in detail about Ayurveda as the kind of treatements available in ancient India and practiced till date is the oldest surviving Medicinal Practices.

Ayurveda is based on the elements that the universe is made up of five elements: air, fire, water, earth and ether. These elements are represented in humans by three "doshas", or energies: **Vata** which indicates nerve impulses, **Pitta** which indicates metabolism in organ and tissue system and **Kapha** the cerebral-spinal fluid that protects the brain and spinal column

- **Kaya-chikitsa**: – GENERAL MEDICINE

- **Kumara-bhrthya**: – PAEDIATRICS

- **Shalya-chikitsa**: – SURGICAL SCIENCE

- *Salakya-tantra*: – ENT

- *Bhuta vidya*: – PSYCHOLOGICAL COUNSELLING basically used to perform EXCORCISM

- *Rasayana-tantra*: – ELIXIR—Combination of Medicines

- *Agada-tantra*: – TOXICOLOGY

- *Vajikarana tantra*: – SEXOLOGY

- *Sushruta Samhita*: – **CATARCT in HUMAN EYE** and is performed with **JABAMUKHI SALAKA** is a curved needle is used to remove cataract

## *THE TREATEMENT PROCEDURES EXPLAINED*

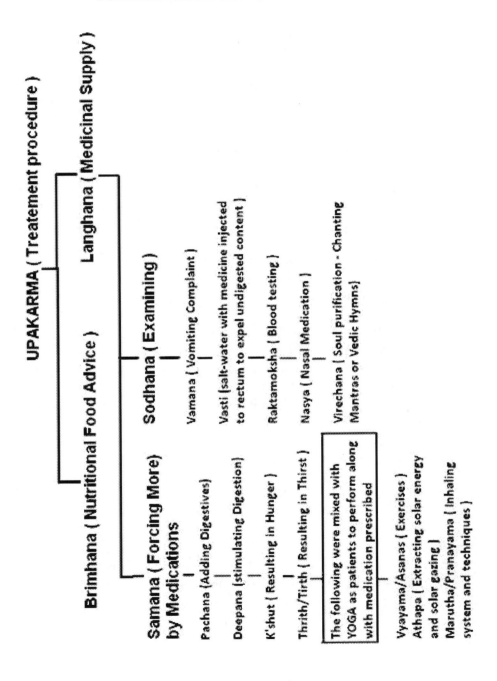

UPAKARMA ( Treatement procedure )

Langhana ( Medicinal Supply )

Brimhana ( Nutritional Food Advice )

Sodhana ( Examining )

Vamana ( Vomiting Complaint )

Vasti (salt-water with medicine injected to rectum to expel undigested content )

Raktamoksha ( Blood testing )

Nasya ( Nasal Medication )

Virechana ( Soul purification - Chanting Mantras or Vedic Hymns)

Samana ( Forcing More) by Medications

Pachana (Adding Digestives)

Deepana (stimulating Digestion)

K'shut ( Resulting in Hunger )

Thrith/Tirth ( Resulting in Thirst )

The following were mixed with YOGA as patients to perform along with medication prescribed

Vyayama/Asanas ( Exercises )

Athapa ( Extracting solar energy and solar gazing )

Marutha/Pranayama ( Inhaling system and techniques )

Dhanurveda—(Warfare stratagies and Weaponry developement)
Gandharvaveda—(Music, Dance, Grammar, erotica)
Arthashasthra—(Public Administration, economy, governance and political science/study)
Stapathyaveda—(Engineering and Technology)

<<<<<<< §µΨΦ∞ЖλπЏδ >>>>>>>

## VEDANGAS

**Vyakarana Vedanga** is Grammar. Panini the grammarian used to define the grammar of Sanskrit in sutras (rules of grammar)

**Chandas Vedanga** : It is the science of distance. Guru-Laghu is measurement called a metre. Metres varied in which the Vedic mantras are composed. It is also used as a tool for error-detecting mechanics. As Vedas were taught generations over generations orally before scripting, any mispronounciation in chandas, error shall be easily identified and corrected.

**Nirukta Vedanga**: It explains the word origination, deriving the meanings of words in different contexts

**Siksha Vedanga**: It deals with phonetics, Udatta and Anudatta the pronunciation & accent consistency.

**Jyotisha Vedanga** : It is Astronomical Science. Detailed study about the movement of stars.

**Kalpa Vedanga**: It is Ritual-Science, the guidelines for poojas and sacrifices is to be performed while performing.

<<<<<<< §µΨΦ∞ЖλπЏδ >>>>>>>

## What we forgot from Indian Civilisation?

Above Notes from the Original Transcript of Vedas giving us the examples of the way of Life which survived from ancient indian practices taught to the world thru TAKSHASHILA UNIVERSITY—World's First University and NALANDA world's first RESIDENTIAL UNIVERSITY

More than 10,500 students from all over the world studied here in TAKSHASHILA UNIVERSITY. The campus accommodated students who came from as far as Babylonia, Greece, Arabia, China and East Asia as per students register records, TAKSHASHILA UNIVERSITY offered over sixty different courses in various field such as science, mathematics, medicine, politics, warfare, astrology, astronomy, music, religion, philosophy etc., Generally, a student enters TAKSHASHILA at the age of sixteen. Students would come to this university and take up education in their chosen subject with their teacher directly. The Vedas and other 18 Arts, which included skills such as archery, hunting, animal science etc.,

Panini (the famous Sanskrit grammarian), Chanakya (Idealogist) and Charaka (the famous physician) and Chandragupta Maurya of ancient India were the proud students graduated from this TAKSHASHILA UNIVERSITY. It gained its importance again during the reign of Kanishka. It was probably, the ancient seats of higher education. Takshashila is perhaps best known because of its association with Chanakya. The famous treatise Arthashastra (Sanskrit for The knowledge of Economics) by Chanakya,

TAKSHASHILA has been listed by the UNESCO as one of the World Heritage Sites.

The period between 600 and 1100 AD is considered popular period of Indian Astronomy as Indian Astronomical science gained popularity because of prominent astronomers like Aryabhatta, Lallacharya, Bhaskaracharya and others. Bhaskaracharya, is regarded as persipicacious almost without question as the greatest ardent Hindu and eminent mathematician of all time.

Bhaskaracharya was head of an astronomical observatory at Ujjain, the leading mathematical centre of ancient India. Some of Bhaskara's contributions to mathematics include the following:

1) Bhaskara is the first to give the general solution to the quadratic equation $ax^2 + bx + c = 0$, the answer being $x = (-b \pm (b^2 - 4ac)^{1/2})/2a$. Solutions of indeterminate quadratic equations (of the type $ax^2 + b = y^2$).

2) He also gives the Trignometry well known results for sin $(a + b)$ and sin $(a - b)$.

3) Bhaskara's deep knowledge into Differential Calculus and suggested the Differential Co-efficient nulls at maximum or minimum value of the function developing concept of Infinitesimals—An indefinitely small quantity or a value approaching to zero.

Bhaskaracharya wrote **Siddhantha Shiromani** in 1150 AD. Significance of **Siddhantha Shiromani is simple methods of calculations from Arithmetic to Astronomy**. Essential knowledge of ancient Indian astronomy can be understood by reading this book. It has exceeded all ancient books on astronomy in India in terms of accuracy and reliability. After Bhaskaracharya, no scholar can write manuscript on mathematics and astronomy in easily understandable version in India.

His work **Bijaganita is effectively a treatise on algebra and contains the following topics:—Positive and negative numbers. Lilavati is an outstanding example of writing a difficult subject like mathematics is written in poetic language**. Lilavati was translated in many languages throughout the world. It consisted the topics of arithmetical and geometrical progressions, terms and definitions, interest computation, plane and solid geometry. Bhaskara's method of solving is easier, than the much complex methods found in the work of Aryabhatta.

His method contributed in **finding out the distance between two places on the same longitude. Then finding out the correct latitudes of those two places and difference between the latitudes. Knowing the distance between two latitudes, the distance that corresponds to 360 degrees can be easily found, which the circumference of the earth** confirmed that earth is round in shape and not flat as we see.

He also calculated a planet is farthest from, or closest to, the Sun, the difference between a planet's actual position and its position according to the equation of the centre which measures positions on the assumption that planets moves at different speed and time around the sun is proved. Therefore he concluded the differential of the equation at the centre is equal to zero. Further led to the calculation of one day in Moon is equivalent to 15 earth-days and one night is also equivalent to 15 earth-days.

Bhaskaracharya had calculated accurately the orbital periods of the Sun and orbital periods of Mercury, Venus, and Mars. There is slight difference between the orbital periods he calculated for Jupiter and Saturn and the corresponding modern calculation. Bhaskara calculated earth's atmosphere extends to 96 kilometers and has seven parts. He accurately predicted that there is a vacuum beyond the Earth's atmosphere.

With our Chapter title, you could understand what we are dealing with is the Indian Idealogies, which once happened in India which moved the scholars all across world providing Indians more to the VANITY of Indian Scholars and their methodologies taught the world of methodical Approach to the inventions and discoveries made against the practical problems.

<<<<<<< §µΨΦ∞Жλπधδ >>>>>>>

I remember of **HU SHIH the famous Chinese philosopher, essayist and diplomat**. Hu is widely recognized today as a key contributor to Chinese liberalism and language reform in his advocacy for the use of written vernacular Chinese. He was one of the leaders of China's New Culture Movement, was a president of Peking University, and **in 1939 was nominated for a Nobel Prize in literature**.

In His Famous Quotes:-

**"India Conquered and dominated China culturally for more than 20 centuries without ever having to send a single soldier across her border"**.

<<<<<<< §µΨΦ∞Жλπधδ >>>>>>>

# 3

# *Electricity of Thoughts*

Many educated men have become very limited perceptioned and pedantic in this world today, rather ended up highly self egoistic, over confidence, have left them to exposed to highly materialistic world for earning money and work hard to get their dreams fulfilled?!!! This journey of mind, full of thoughts is directly addressed to the above mentioned assimilated men and women in world today.

As the world have become, money centric and as Indians did we ever thought it was our forefathers have taught Metaphysics and Quantum physics thru the epic stories and we all remember that our elders told us the bedtime stories about the different realm of the worlds, heavens and the visits of living beings from different worlds and heavens. Gods from heavens visitied earth to save humanity and righteousness, which was very high concepts in spiritual physics, in the subject matter of spiritual science. But what we do in inturn giving our young kids is the concepts of material world, Today's kids were driven more towards eateries, cinema, fashion and school education particularly had become more marks or grade centric for creating competition, which in turn developed by-hearting the school lessons rather understanding the concepts in the lessons.

I would rather say or put it in words that we are driving younger generations delusional towards the materialistic result oriented competitive environments, may continue till the same kid starts to think what is that he or she needs to compete for?, how long? for what results? The formula on goal setting and achieving the goals is what we teach kids? what we are not teaching them is problem solving skills and analytical approach? Indian elders taught this in the bedtime stories of Lord Krishna and Lord Rama about how they took the help of environment around and how they stood for righteousness and how they defeated their enemies by identifying their weakness and making right moves wisely thru smart work.

Perhaps this generation parents teaching the same to their kids or they simply pushing their kids towards goal settings?? Perhaps in first place the parents of this generation could atleast tell a story to their kids with the essence of the concepts and make the kids understand it (or) is it going to be a story telling for a mere story telling??

If a parent thinks that he is doing the justice for a kid? as indians wouldn't have lost YOGA in their daily life and would have not driven our kids towards junk foods like pizzas, burgers and fried chickens? Natural Herbal shikakai and turmeric would have not been replaced by highly concentrated shampoos and body creams which only made us to search for DERMATOLOGISTS?? the most an Indian should have been ashamed of is that we Indians cultivated the Turmeric crops, used it in cooking and in medicines have sacrificed the patent rights to a country which doesn't even had the proper climate of growing it? How many number of herbal patent rights we have sacrificed to how many countries??

<<<<<<< §µΨΦ∞Ж入πЏδ >>>>>>>

How many parents of this generations knows the simple basic exercises of YOGA? Geneuinely? 15 % in today's India? western countries, now started adopting the India YOGA for meditational Practices? Pranayama (Breathing Techniques) and Aasanas (Exercise Postures)? Do we atleast teach it to the kids in Schools as co-curricular activities? How many Schools today? which made me look on Chinese how they inducted the QI GONG into the school education as one of the subject like Moral science and general knowledge QI GONG is Chinese Yoga, conceptually the Indian Yoga and Chinese Yoga were almost same. It's the perfect injustice we were doing to the young india? A parent cannot compensate his injustice only by spending more towards the kids education and entertainment.

For that matter, the more we forget of cultural heritages and compromising it for the western suggestions and ideologies the more American dollar and euro making its way into india and "Iam not the protectionist and at the same time iam not the one who is willing to let go everything", If Dollar and Euro entering india in the name of investments and business the more india have become the profit generator for dollar and euro, thanks to Indian politicians who made india dependent on Foreign Institutional Investors and Investments, happening so,

I think India should open the doors for trade and commerce for agricultural products and spend more towards the research in agriculture and farming, rather wasting money on giving freebees and discounting loans issued and stop accumulating bad debts, leading the nation to current account deficits to its record heights and leaving the GDP to its lower levels, I strangely feel its funny enough to see things happening inside out, What do you feel?? If this is what Harvard—Business administration and Finance degree and Ph.d in economics can do to india? what is the use of those degrees sitting in political governance of India?

Sadly, I say we are loosing india left, center and right, by giving it to the greedy minds? How do we streamline this massive whirlpool of western culture is taking out india? I would say by properly propogating it into the minds of money who commercialises and brand merchandises the indian culture but one should understand that branding the Indian culture is not in sing india sing not in music alone or not dance india dance and not in dance alone? We have it in Ayurveda and Herbal Medicine, YOGA, Agiculture techniques and systems, Modernising the concepts of Ancient Engineering and Technology, Indian Banking, Trade and Commerce techniques, skills oriented handicrafts, arts and painting.

<<<<<<< §µΨΦ∞Жλπμδ >>>>>>>

Many Indians who revolutionized the modern day computer applications, one Mr. V.A. Shiva Ayyadurai a mumbai born south Indian developed a full-scale emulation of the interoffice mail system, which he called "EMAIL" and copyrighted in 1982. Ayyadurai teaches Systems Visualization at MIT. In 2012, he launched Systems Health, an educational program for medical doctors which integrates concepts from systems of holistic medicine such as Siddha, Ayurveda, and Traditional Chinese medicine with systems science and systems biology. Systems Health is offered through the Chopra Center with Deepak Chopra, a holistic health/New Age guru and perhaps the most famous of America's alternative medicine practitioners. Why such minds were not accommodated in india? Why Indian governments were not aware of such skills and why india is not spending on such minds which is capable revolutionizing the nation. Does India don't have the money and infrastrucuture to support such capabilities? I would say Indians as a whole and every single citizen to be blamed for not giving such oppurtunities to somebody who has visions to foresee the availabilities.

<<<<<<< §µΨΦ∞Жλπμδ >>>>>>>

Any Indian have ever thought that the Currency trading in Futures exchange is illogical practice to have the currency manipulation, Economists say, Currency

Futures is the best parameter for the currency to determine its fluctuating values, I ask HOW? A FOREX market is 24 hour traded market, When world currencies were traded now 24 hours, it is implied forex market is 24 hours market.

A Commodity can be future traded for example an agro commodity could be future traded as farmers and traders tend to fix the future price for the commodity due to unwarranted monsoon and climatic changes, perfectly logical, for that matter, even shares will do fine?!! good and logically sound enough. How come currency trading in futures exchange?! The changing future price will be fixed by changing currency future value? it is only a manipulative tool used by banking and financial system to fluctuate the currency, or I ask the economists what is the need of currency to be traded in futures exchange??

Currency futures have been introduced by Chicago Mercantile Exchange for determining USA $ value fluctuatuions and settlements against gold standard rate systems but later suspended against gold standard rate as settlement rate. Now we have allowed it to value against other currencies and the currency value we derive is actually floating value not measured in any parameters against any settlement value. Trading commodities against commodities is a Barter system and is not allowed legally in exchanges? Trading currency against currency is not barter system? and is allowed legally in exchanges? How? when currency itself has become a commodity in trading platform in futures traded? is there any sound logic making it logical??

Speaking of Logic US $ in currency value, lets say US $ index is valued against six major currencies holding the share of 1 US $ to Euro 57.6%, Japanese yen 13.6%, Pound sterling 11.9 %, Canadian dollar 9.1%, Swedish krona 4.2%, swiss franc 3.6%.

So how does it become logical to Indian Rupee to trade against 6 currency value in the name of US $, lets say Euro, How does Euro's 57.6 % Trading against 43% US $ value again, because US $ holding 57 % value from Euro's Value? So If 1 Euro is spent to buy 1 US$ it means buying 0.43% of US $ by spending 1 Euro? any logic?? It is approximately averaging between 1.32 Euro required to be spent to buy 1 US $, The excess of 0.32 euro is spent to towards the premium for the loss. Europeans seems to be loosing more than anybody? Pity on Europeans?!!!!

European Union is awarded a Nobel prize *for over six decades contribution to the advancement of peace and reconciliation, democracy and human rights in Europe".*?? **Financial stability is lost to gain peace, democracy and human**

**rights**? so if you trade US $ in forex it is implied you are trading against 6 national currency value in split percentages? This is how a currency valued??? American $ is clearly faultering the world economy by injecting an unexplainable wrong trading concept called forex futures trading. How do you explain this?

Let me put it this way, A tennis player playing his game with multiple opponents and trying to score his points to win? How does he score a point to win? or is it the logic for the game to be kept played continuously without scoring a point or it is simply controlling the unfinishing game with audience expected to bet on and acknowledge it as legal,

**Rather currency futures value could be determined by allowing only the central banks across the world to trade it against its Current year GDP, Current Account Deficits, Inflation index, Outstanding government loans borrowed and lended, forex reserve Index and Bullion reserve Index as cumulative settlement values for parameter to determine the currency futures value.**

**what will happen, Each country will be measuring the US $ value on daily basis against its currency at different levels and the average of that value will become the US $ currency futures value, the fluctuation in US $ currency value will be arrested and US $ may fail to sustain its status as world currency?!! the forex futures trading is introduced to destabilize the competing economies. Economists and THINK TANKS should do their job firmly and fairly now.**

Currency in futures exchange is a CANCER to any nation, many countries in the world have not opened the currency futures trading, and not willing to do as well, without thinking twice Currency futures trading to be suspended. It is national interest to decide whether to allow it or suspend it. India as an emerging market is not competing with leading economies and trying to involve in digital trading systems only by opening its gates for currency valuation against a developed nation? thru currency futures trading? No it was very wrong perception.

**Why America could not save its economy while having its currency as world trading and travelling currency? Economists please think? Indians please think? Politicians please think? Why America got into the concept of Quantitative Easing? Bond buying programmes, Pumping in Money, Printing Dollars and injecting into economy? Do you know how they measured their Quantitative easing techniques is effective?? by the reduction in unemployement, even mighty strong american dollar in value could not save their economy,**

35

currency futures trading did not helped the economy it was printing American dollars, the real money not the American dollar's value in currency furtures exchange?!!!!!!!!!!!.

I remember Mr.Bernanke's exact words in a conversation at NBER in the words of Mr.Bernanke . . . . "Well, I gave some remarks on this at a London event for Mervyn King's retirement and appropriate of today's discussion, I used historical examples. I made a distinction of during the 1930s, during the Great Depression, as countries left the gold standard, their currencies temporarily depreciated relative to other countries, and they had a temporary trade advantage because of that; but over time, as all the countries left the gold standard, exchange rates kind of normalized, kind of went back to where they started from, but nevertheless the whole world was nevertheless much better off because there was a global monetary expansion which was desperately needed at that time, in the 1930s".

I don't know Mr.Bernake realized this, let me take the opportunity of explaining this, when economies of developed nations were suffering in 2008-2009 collapse many asian countries just shyed away recession, How? It was gold came in handy to help the asian economies? It was gold that central banks of asian economies reserving in their lockers?

I put forth addressing these problems because the majority of the indian population doesn't know what these issues are about in first place. I don't want my brothers and sisters to pay their hard earned money to the illogical trading concepts of currency futures. **Any futures exchange should speculate value against asset it is logical. The Younger generations were watching us if we start speculating the value against value, this world is going to be gambling world, Imagine a world of futures exchange of family sentiment and values, a husband can trade his son or daughter or wife's beauty, intelligence, health or her maternal capabilities. Could you agree on it being future traded?? its so simple to adopt or discontinue before it destroys you.**

<<<<<<< §µΨΦ∞Жλπ∐δ >>>>>>>

Now someone, lets say Mr. Financial expert is giving imperative advise to Indians not to buy GOLD?? Trading of gold coins and biscuits have been arrested?? for what?? because they fear that currency value will go down? Mr. Financial Expert says it was gold which is saving Indian rupee from falling further and Mr. Financial expert thinks that discounting financial loans for a massive 1,20,000 crores did not affected economy and 2G spectrum scam did not affected economy?? Mr. Financial expert says the gold, the physical commodity is downgrading currency value?? Indians were not fools to have gold for their security, They were smart enough to know when people like Mr.Financial expert is doing disaster mistakes by adopting the illogical currency derivatives concept introduced and announcing loans discount rather Mr. financial expert should think of increasing energy resources, research funding for the development of soil fertility and protection techniques and agricultural products and water conservation projects to help the agricultural lands and farmers.

<<<<<<< §µΨΦ∞Жλπʮδ >>>>>>>

Educational systems broadly in india requires a lot of changes, Schools had 7 period till early 1990's A period of 40 minutes session x 7, later it has been extended to 8 hours x 50 minutes. Technically 120 minutes 2 hours of school timings has been increased? Why? is it due to a complaining parent that they could not have their kid managed at home? What do we teach? what the extra skills do we receive from schools? is there 1 period a day in developing multi skills. A separate government regulatory body is required to set up to propose and oversee the operational guidelines of developement of indoor extra curricular lessons being fulfilled by licensed schools in india.

Basic YOGA shall be proposed and to be mandated in school level education at least till the primary level till fifth grade, will make the Indian kids to get enriched Indian physical exercise system conceptually as well as real time training.

Why not this idea to be putforth by the parents to the schools during open sessions or from Parents teachers association. Such suggestion implemented shall become a revolutionary system into schools and it should be brought with subject code by giving separate theoretical and practical exams and should be mandated on passing minimum.

<<<<<<< §µΨΦ∞Жλπʮδ >>>>>>>

The mind should not be allowed to have practiced to believe what the eyes see and what ears see, How an average salaried Indian lost his money during 2005-2010 any salaried middle class men and women will have their life insured in life insurance policy. **After newly elected government in 2004 took over governing office the finance ministry of india decides to revolutionise the Insurance industry,** introduced the new insurance concepts called **ULIP (Unit linked Insurance Plan)** which means the premium amount you pay will be invested partially and systematically in equity, debt and bond markets, Anyone should not forget that the ULIP in insurance concepts is failure, India launching a failure model, the so called Indian corporates engulfes it and celebrates it by advertising enormous returns, **with media advertisements ends "ULIP is the subject matter of risks and market solicitation"** and **customer is advised to read the documents carefully before investing"**, <u>investing yes it said investing not purchasing</u>, **the very fundamental meaning of insurance has been killed** and the concept of mutual funds brought into the concept of Insurance making people to misconceive, **as the investment will be protected under insurance companies unlike the mutual funds and demolishing the mutual funds industry to failure. Great it happens only in india and it happened.**

I knew by 2005 when markets selling ULIP under insurance companies, I already knew Indians going to loose hard earned money and it happened as well when 2008 world recession hung on mountain it happened Insurance companies sold ULIP went into loss and could not even get its market to reach unit purchased value, Indians lost money, companies went into loss and everybody struck in solar eclipse effect, Government dont really know what to do and they formally requested their customers that 3 years lock in period is much less and pleaded for 5 years lock in period and changed it by passing a bill. Where do insurance companies stand now? Insurance companies one by one slowly stepped back from selling ULIP and got back to selling of traditional conventional insurance plans and many companies have suspended their ULIP plans altogether.

<<<<<<< §μΨΦ∞Ж&lambda;πЏδ >>>>>>>

Reasons for highlighting these problems and failures is India facing in last 2 decades, once India taught the world what the cosmological science and mathematics and chemical science is what india have become, reasoning due to casual attitude the people were developing and as an assimilated society it reflects Indians started forgetting what we had and what is our strength is . . . .

How do we regain and stabilize ourselves from the fall which is happening? any new business model introduced under privatized sector by indian government to be left for advisory panels which consists of all sectors of society to be reviewed and well analysed atleast for an year, the practise will have its potential to become the key factor of anti corruption, A special ordinance could be passed only after the advisory panel members approving the idea by 85 % minimum to qualify for introduction, after all state political parties agreed upon with a special jury consisting all sectors of society acknowledging it. A proper systemised political operational structure alone will solve the future issues before uprising.

<<<<<<< §μΨΦ∞Ж λπЦδ >>>>>>>

What do teenagers and youngsters is doing now is toying their mobile phones, i was wondering what mobile phones and its applications have done to our youth is to made them to listen to music holding them back from looking into valuble informations which could turn them to paradigm shifts and even worst it made them to play games in their mobile phone, completely psyched them out, from converting their idle time to convertible values. I see mobile phones rather to be called smart phones whether it is fulfilling the needs of communication needs I really don't know . . . . but its success displayed in ensuring as it is serving the entertainment needs of the customer.

I suggest the usage of powerful internet tool as the life changer is rather being under utilized, rather shows the success of porn websites hit counters increasing to peak to establish the flags upon internet. The upcoming younger generations life will change if they start spending time in watching some fantastic discovery networks, National geographic and BBC documentaries available over you tube, rather wasting time in viewing movie trailers. Please do explore internet about infinite available information and oppurtunities to be materialized. I knew a friend who learned cooking thru internet videos have utilized the ideas and have become a successful chain restaurant owner. I agree to the famous quotes "count your chickens before they hatch" a true success mantra.

<<<<<<< §μΨΦ∞Ж λπЦδ >>>>>>>

As Indians really need to understand a simple fact "Don't trust everything blindly which is being displayed in front of you" Do analyse I will give a fantastic example It was Febraury 8, 1930 we have been reading as Pluto as planet of our solar family, With Eris being larger, made of the same ice/rock mixture, and more massive than Pluto, the concept that we have nine planets in the Solar System began to get aquainted and redundant.

What is Eris, planet or Kuiper Belt Object, what is Pluto, for that matter? Astronomers impelled they would make a final decision about the definition of a planet at the XXVIth General Assembly of the International Astronomical Union, which was held from August 14 to August 25, 2006 in Prague, Czech Republic. Astronomers from the association were given the opportunity to vote on the definition of planets. One version of the definition would have actually went ahead by discerning the number of planets to 12. Pluto was still a planet, and so were Eris and even Ceres, which had been thought of as the largest asteroid. A different proposal kept the total at 9, defining the planets as just the familiar ones we know without any scientific rationale, and a third would drop the number of planets down to 8, and Pluto would be out of the planet count. So . . . . what is Pluto?

<<<<<<< §µΨΦ∞ЖλπЏδ >>>>>>>

**That's what modern science is doing to our kids in their school lessons and to our knowledge. Modern science officially announces confirms something today and counterfeit it tomorrow and gets convinced and celebrate new theory the next day unlike the PERPETUAL VEDAS.** when we start fervent in this modern day science discoveries particularly pertaining to Astro physics, the so called Space Research organization were not discovering anything new in space?! How something could be discovered when it is not kept hidden, these planets and stars were already there out in open space. These modern scientists claim they have found something, using a telescope, which they claim its powerful enough to find new stars, has its limited capacities, the very terms they were using "new stars found" how they say it is new star as if it was lost yesterday and later announced found today, these things were not joke really, and to the funniest of all is that these scientists(Space Researchers) keeps their name to the stars as they claim to be found by them.

Atlast normal people like us don't know the subject matter of physics to its deep willing to hear the collision of two great galaxies as we know it as BIG BANG (theory) and super nova forming the new and parallel universes, and they say this event of our universe belongs to the galaxy thank god modern scientists and astronomers didn't named the galaxy behind their names and still couldn't give the precise time periods, ok excused because is understandable as space time is different from earth time.

<<<<<<< §µΨΦ∞ЖλπЏδ >>>>>>>

The journey of the mind begins here to allegorize and find how impelling the illusionary world moving the real world and facts to shake its foundation, Now the problem being analysed here is the matter difference between the illusion and reality, The illusionary world often have a practice to convince the real world, is they mix up a percentage of lies or hide the facts to the real world fact, which most of us not aware of what is hidden, its very simple point to understand in the world of black and white there is always the grey area where the trap is set to falter the real world anamoly, where the majority sector of population set to either get convinced on this deviation or failing to understand the deviation will end up as listeners and loosers. The well planned designers of deviation being the tellers and winners, It is absolutely no way any teller can be tellers always, certainly they will be questioned on the deviated facts will be always become answerable and liable to their designs. It should be listeners responsibility to safeguard their interests in the process of understanding the deviations and stand to what is in the limits justifiable borders. Change is always to be welcomed and embraced but one should be careful about the change sticks to fundamentals of reality. Remember when there is no loosers? What about winners????

I remember of fitting quote by Lao Tsu, a Chinese philosopher who is considered as the father of Taoism quotes "Truthful words are not beautiful and the beautiful words are not truthful. Good words are not persuasive and persuasive words are not good". So on this what answer do we have? Again Lao Tsu quotes "Mastering others is strength, mastering yourself is true power".

It should be us in our considerable thinking that we should able to understand the self and environment and society around us to make our stands to stand to decide what you want and what you should discard considering the facts analysed, the more emphasis given on the thought process is itself becoming the journey of mind dwelling into the pool of electric thoughts a second, arising in mind.

<<<<<<< §µΨΦ∞Жλπμδ >>>>>>>

We need to dig little deeper to it by understanding the largest spoken language of asia and sharing the similarities in other languages.

**Addressed to the Asiatic Society by Sir William Jones (1786)** . . . .

"The Sanskrit language, whatever be its antiquity, is of a wonderful structure, more perfect than the Greek, more copious than the Latin, and more exquisitely refined than either, yet bearing to both of them a stronger affinity, both in the roots of verbs and the forms of grammar".

Sanskrit (Literary language), a language spoken in ancient India, is part of the Indo-European language family. As the name suggests, this family includes Sanskrit and its descendants along with most languages spoken in Europe, Southwest Asia and central Asia. Sanskrit, Latin, and ancient Greek form a trio of classical literary languages. Sanskrit is the ancient language of India and the Indian subcontinent. Its literature, the Vedas, was written in Vedic Sanskrit, as in the *Rig Veda*, Chronologically next came Classical Sanskrit 4th century B.C. The literature giving the evidences of Vedic science much before 6000 years from now.

A family of languages (including most of the languages spoken in Europe, India, and Iran) descended from a common tongue spoken in the third millennium B.C. by an agricultural people originating in southeastern Europe. Branches of Indo-European (IE) include Indo-Iranian (Sanskrit and the Iranian languages), Greek, Italic (Latin and related languages), Celtic, Germanic (which includes English), Armenian, Balto-Slavic, Albanian, Anatolian, and Tocharian. One should remember the time periods we are describing is when Middle east countries devoid from Judaism, Islam or Christianity as religion.

Words that relate to families are also similar in most Indo-European languages

| "FATHER" | "BROTHER" |
| --- | --- |
| pitar (Sanskrit) | bhratar (Sanskrit) |
| pater (Latin) | frater (Latin) |
| pater (Greek) | phrater (Greek) |
| padre (Spanish) | frere (French) |
| pere (French) | brother (Modern English) |
| father (English) | brothor (Saxon) |
| fadar (Gothic) | bruder (German) |
| fadir (Old Norse) | broeder (Dutch) |
| vader (German) | bratu (Old Slavic) |
| athir (Old Irish—with loss of original consonant) | brathair (Old Irish) |

Similar words found in Sanskrit and Latin compared

| MEANING and in ENGLISH | SANSKRIT | LATIN |
|---|---|---|
| Three | Trayas | Tres |
| Seven | Sapta | Septum |
| Eight | Ashta | Octo |
| Nine | Nava | Novem |
| Snake | Sarpa | Serpens |
| King | Raja | Regem |
| Gods | Devas | Divus |

Being astute and erudite what the world is digesting today, lets contemplate it in the proceeding chapters to see more about what and how Vedas tried to teach us and how we profound the modern science and remember it now. It always happened modern science tried to prove the Vedas as merely a myth and failed concept of truth, later understanding the depth of Vedas, modern science started to taking the lead from Vedas and it is actually the Vedas helped modern science and scientists and physicists to understand the conceptual universe from since, then and now . . . . We explore a little more how Ancient India's philosophies travelled the other parts of the world and religions renowned its value and became the part of the society and bridged its way in the next chapter . . . .

<<<<<<< §µΨΦ∞Ж λπЏδ >>>>>>>

# 4

# *History Repeating itself in Different Dimensions*

After addressing the modern day communities and societies of casual attitude of many nations, we continue our journey of thoughts towards ancient era of science and modern science, In the 1st chapter we tried to understand the Ancient Indian wisdom and philosophies grew into one of the most scientific religion in the world. In this chapter we ride into a path and see how primitive ages gave birth to civilizations and the similarities found in them

Ever since the civilisations started in the primitive world before the Pre Modern Day religions, Primitive people lived in the dark days used to live in caves were extremely afraid of nature due to lack of intelligence and in difficulty of understanding the changes of nature, After generations passed when man started to understand his environment inventing fire, tools to move around and the civilisation gradually started growing,

**So What is civilization basically?? Civilisation is humans, understanding the needs of essential supplies like food, water and shelters, decides to make shelters where the abundant water resources found, and nurturing himself and his dependents with the set of governing practices imposed on with an idealized divine belief system.**

Thru out the history, **The rise and fall of civilisation happened due to the excess or scarcity** in other words, When more rains poured and flooded the river or due to no rains, civilisation suffered with floods, drought, famine and poverty and the land of civilization collapsed which led the inhabitants to migrate to different land areas in search of water. The Great Egyptian, Indus valley, Mesapotomian and other asian Civilisations were all river based civilisations.

The other civilisations which grew is in the high altitude regions, mountain civilizations, This civilization chose the mountain and hill regions for the same reason with one advantage, water and vegetations kept the mountains green to its height it became extremely difficult to the intruders to attack or dominate, this left them safe on high grounds and secured state of living. It was found this civilization mostly found in North European Parts and North Western Parts of Russia (Scandinavian region) were called Vikings known today identified with Norse Religion and followed Old Norse Mythology, Similarly the Aztec civilization in Mexico and Andean Civilisation from North, Central and Costal Mountain ranges of Peru, In those days these civilizations were not geographically seprated countries with borders.

Towns and Cities were built, as more and more humans and cattle inhabitants along with the migrants started living closely as a community understood the essence of living together and need to defend and help each other from the external dangers, lived near huge rivers which ran across the massive land in the midst of the great oceans. When civilisations grew, the wisest who lived among the community gained the momentum to lead the civilizations grow to greater heights, which led to the trade and commercial routes connecting to the communities together lived across the land in different places, when people among themselves chose their leaders and pronounced them as their kings, trade and commerce grew, cities grew, kingdoms grew which attracted more and more migrants towards civilizations.

**The following world map will be lucid to comprehend the Ancient civilization.**

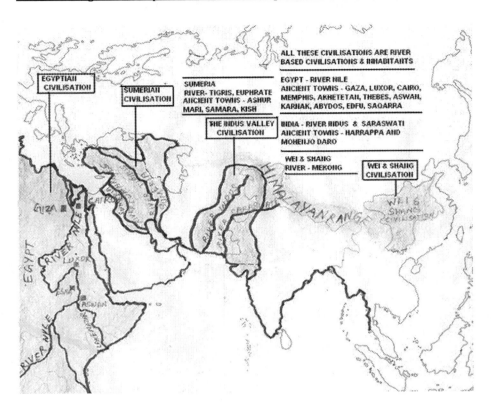

On a very Intersting Note, Similarities found in Myths and Epics stories mention about gods told in the above mentioned civilisations.

## COSMOLOGICAL IDEA OF ANCIENT CIVLISATION

| SUMERIAN CIVLISATION | EGYPTIAN CIVILISATION | INDIAN CIVILISATION | MODERN THOUGHTS |
|---|---|---|---|
| **The Sumerians envisioned the universe as a closed dome surrounded by a primordial saltwater sea.** Underneath the terrestrial earth, which formed the base of the dome, existed an underworld and a freshwater ocean called the Apsû. The deity of the dome-shaped firmament was named An; the earth was named Ki. First the underground world was believed to be an extension of the **goddess Ki, but later developed into the concept of Kigal. The primordial saltwater sea was named Nammu.** CONCEPT OF (TRINITY FOUND) | It was the fixed, eternal order of the universe, both in the cosmos and in human society. **In Egyptian belief, Ma'at was constantly under threat from the forces of disorder,** so all of society was required to maintain it. On the human level this meant that all members of society should cooperate and coexist; on the cosmic level it meant that all of the forces of nature—the gods— should continue to function in balance. The **Egyptians also believed in a place called the Duat, a mysterious region associated with death and rebirth** CONCEPT OF (TRINITY FOUND) | The Rig Veda's view of the cosmos also sees one true divine principle self-projecting as the divine word, *Vaak*, 'birthing' the cosmos that we know, from the Hiranyagarbha or Golden Womb. The **Hiranyagarbha is alternatively viewed as Brahma,** the creator who was in turn created by God, The universe is constantly expanding since creation and an alternate view is that the universe begins to contract after reaching its maximum expansion limits HINDU Cosmolgy speaks about both **Mathsya avatar and huge ocean engulfing mentioned in Sumerian & Greek Civilisation.**The idea of Atman and Paramatman and the nature gods Rudra mentioned in Egyptian Civilisation **CONCEPT OF (TRINITY FOUND)** | **The Hindu cosmology and timeline is the closest to modern scientific timelines and even more which might indicate that the Big Bang is not the beginning of everything but just the start of the present cycle preceded by an infinite number of universes and to be followed by another infinite number of universes.** It also includes an infinite number of universes at one given time. |

The following table of the Gods of ancient civilization and incisive idealogy behind

| SUMERIAN GODS | EGYPTIAN GODS AMUN (Creator) | HINDU GODS NARAYANA (CREATOR) | GREEK GODS |
|---|---|---|---|
| (TRINITY) + 4 = 7 **Anu** (Lord of Heaven) **Enlil** (Lord of Storm) **Nenlil**(consort of Enlil) **KI**(Godess of fertility) **Enki** (Patron City God) Ishtar(Godess of love, War and Sex) **Sin** (God of Moon) shamash (Sun God) | **RA (SUN God) PTAH (Body of Ra)—Planets of RA's Solar Family Sekhmut (god of Destruction) ISIS (God of MAGIC) THOTH (Wisdom God)** | Trinity—Vishnu (Form of Creator NARAYANA) the gravitational force pulling for Galactical explosions and BIG BANG organiser and the form presenting himself in other worlds. **BRAHMA (the Creator of Universe) Scientifically and SUPERNOVA the erupted Polaris star (Western) and Dhuruva (Hinduism) the North pole star. RUDRA (The Destructor) and much more to than 100 Demigods. Godess Lakshmi—Wealth Godess Parvati—Health Godess Saraswati—Wisdom** As more and more civilizations grew a lot of subsect in Hinduism and people identified, formed and followed the religious saints and Demigods way of life | Cosmogony (the creation of the world), **Deucalion's Flood** (the re-cycle) parallel to Matsya avatar of Hinduism) **Zeus** (King of Gods— Rain and Thunder) **Hera** (wife of Zeus) God of wedding **Athena** (Godess of Cunning Wisdom) **Apollo** (God of Sun, arts and Healing) **Demeter**—Godess of Fertility & Agriculture **Poseidon**—God of Sea & Horses **Aphrodite**— Godess of Love & Beauty) **Hermes**—God of Travel, Business, weights & Sports. **Artemis**—Godess of Childbirth, hunting & Archery **Hephiastos**— God of Volcano, fire and Craftworkers. **Dionysos**—God of Emotion & Wine. |

| | | |
|---|---|---|
| | | in their own to follow their ideas and principles, As far is any subsect the Concept TRI GODS remained the same in all Castes and Subsects, Framing the stronger Roots to Religion. |

None of the above civilisation's scripture or Primitive paintings or any archeological evidences proved the above civilizations were all dates ranging from 5000 BC to 900 BC calculated by the limited resource, knowledge and tools available to archealogists.

*How can i ignore Hamanuptra? if i miss it would be a complete injustice for a rendition*

Hamanuptra, is the name sounds familiar? Thanks to Hollywood again, Re-popularising it in the movie series of "The Mummy" as Hamanuptra is portrayed as the "City of Dead". Originally the actual Hamnuptra was found in India. During the mid 18th century when India is under british rule, the british engineers were constructing the railroute, and the lost site of Hamanuptra was found, on doubts the work has been halted and the archeological survey team has been sent for further examinations, after the years of excavation the site was identified to be Hamanuptra, The materials found during the excavation further confirmed the cultural connection with Mesapotomian Civilisation.

***Clarification on the Controversy:*** *the Hurrian Civilisation actually predating Egyptian civilisation*

During the 18th Dynasty in THEBES of Egyptian civilisation extended approximately upto 300 years as per historical records, research on evidences confirms several diplomatic marriages happened with the parallel civilizations for strengthening the external affairs during that period, which further confirming the new kingdom starting from Egyptian king of Thutmose III's queen's name satiya (a famous Indian phonetic name) and continued with Egyptian King of Thutmose IV marrying to

Mutemwiya daughter of King Artatama I King of Hurrian of Mittani—Land of Khabur river valley, believed to be spoken of Indo European language possibly old Sanskrit or Pali, Pali is the language spoken during the period of Lord Rama's incarnation, which is also being confirmed in the Indian Ithihasa (Epic Story) THE RAMAYANA. Hurrians were closely related to the land of Isana (Lord shiva),

Wrongly spelled in anthology as its not Hurrians it is to be considered as "Harians" or "Aryans". Hurrian Religion is one of the old religion name of India (Bharatha kanda—major part of asia) during the Egyptian dynasties confirming the time scale of Lord Shiva and Lord Rama predating the Egyptian civilization, the historical Egyptian and Sumerian myth Lord ENLIL (the god of storm) and (NINLIL—Enlil's Consort) is in exact match and co-ordination in epic details of Lord shiva (the God of Storm—considered with the qualities invoked of RUDRA) and parvati from Hurrian religion admonish is still predating in Treta yuga as per Indian Vedic text, The Egyptian Text confirms that Lord Enlil made of fragrant bed of Cedar forest, made love to his wife Ninlil and the consort had encouragement of great pleasure having sex. Lord ENLIL realized that Ninlil was the woman born for his realization and the meaning for ENLIL's life and made her sat on his left thigh and made people to worship NINLIL as the DIVINE CONSORT of Lord ENLIL, The same epic events found in shiva purana Lord Shiva and Godess Parvati of their togetherness is conceptualized in the name of ARTHANAAREESWARA in Hinduism. Lord ENLIL determined the fate of Lady ARURU proud woman surpassing the mountains was considered as one of the most known consort of Lord ENLIL, Lady ARURU is in exact match of GANGA as RIVER GANGES the godess of vegetation offering fertility to the land of India in Hinduism and earlier in hurrian, It was the river ganga surpassing the mountains of Himalayas.

Enlil teaches his son, the war stratagies to Ninruta (Warrior God), is Lord Karthikeya (in Hurrian/Hinduism) teaching him on a strategy to slay the Demon Asag, the epic events which is exactly matching to the Lord Shiva teaching his son Lord Karthikeya trained to kill the Demon Tharakasura, Sharur(in Egyptian myth) a powerful weapon could fly distance and in great speed. The myth of sharur gets its complete treatement as Trishul (in Hurrian/Hinduism), Lord ninruta and lord karthikeya is closely identified with the symbol of fighting cock confirming both in Egyptian and Hurrian civilization.

<<<<<< §μΨΦ∞ЖλπЦδ >>>>>>

Egyptian 19ᵗʰ Dynasty extending to New Kingdoms as Hurrian rulers dominating the Egyptian civilization, The Hurrians gradually gained their international political power and extended their influence till Syria, The King Ramesses I was disapprobation with Lord Rama (Hurrian/Hinduism), No?! King Ramesses I was identified to be the son of commander in chief named SETI (Indian phonetic name) newly formed dynasty of Hurrians, His uncle Khaemwaset, an army officer married **Tamwadjesy, the matron of the Harem of Amun**, who was a relative of Huy, the Viceroy of KUSH, an imperative designation. This confirms the influential status of Ramesses family. Ramesses I found favor with Horemheb, the last pharaoh of the 18ᵗʰ Dynasty.

King Ramesses I was named after the very popular and favourite deity of Hittie and Hurrian religions by then is Lord Rama son of King Dasaratha, Lord Rama is considered as the intellectual king a divine incarnation of Lord Narayana holding the highest rank in Hindu Deity till date.

Hurrians were keen to see how king Ramesses I of his astute and erudite in nature was closely identified of the same abilities of Lord Rama of Treta Yuga in India much long ago. People of 19ᵗʰ dynasty in Egyptian civilization believed King Ramesses I is the incarnation of Lord Rama, as dynasty was flourishing all over middle east civilizations the discerning style of administration of King Ramesses I reflected closely of Lord Rama, not surprisingly Lord Rama being the ancestors of King Ramesses I from Hetite and Hurrian origins.

The royal queen of King Ramesses I carried the name of Sit'r again phonetically sounded the Queen Sita consort of Lord Rama in Hinduism, significantly proving the popularity of Lord Rama, the 7ᵗʰ incarnation of Lord Narayana, clearly considered as the most awaited and upon birth most celebrated after his life time all over the world continued during Egyptian civilization to till date,

As per Shiva(Enlil—Ancient Egyptian) purana, clears with further clarification Godess Parvati or Ninlil (Ancient Egyptian) enquires the purpose why Lord Shiva or Enlil is always penancing and meditating most of his time, on doubts she enquires for realizing what, Lord involves in this act, replies Shiva he is always thinking about Lord Rama and awaiting for his arrival the incarnation of Lord Narayana/Vishnu.

Long before these civilizations there lived people in primitive age secured the proof in one of the Greatest Itihasa or Epic from Bharatha kanda or modern day India, Scriptures dating back to Apeman and the early civilized man lived in same period of time is mentioned in scriptures, Hanuman, bali and sugriva (were all apemen) helping Lord Rama as per the scripture of Ramayana. For everybody's surprise found mention of Huge Aircrafts called (VIMANAS) which floated above the cities during war time. In Vedas to built an aircraft the metals and measurents and chemicals were used, even mentioned in Vimanika Shastra.

The Ramayana remained epic stories to many asian countries outside of India, The original Valmiki version has been adapted or translated into various regional languages, which have often been marked more or less by plot twists and thematic adaptations.

| Countries | Ramayana in their Languages |
|---|---|
| Sri lanka | Ramayana |
| Thailand | Ramakien |
| Cambodia | Reamker |
| Burma | Yama Zatdaw |
| China & Tibet | Ramayana epic found in Dunhuang |
| Laos | Phra Lak Phra Ram |
| Nepal | Siddhi Ramayan |
| Java, Indonesia | Kakawin Ramayana |
| Malaysia | Hikayat Seri Rama |

**A Battle Scene from the Reamker: Rama and Ravana performed at Silver Pagoda, late 1920's**

The tales of the Ramakien(Thailand) is the same of the Ramayana some of the characters linguistically pronounced and spelled in Thai.

- Phra Narai/Witsanu (Narayana/Vishnu)
- Phra Phrom (Brahma)
- Phra In (Indra)
- Phra Isuan/Siwa (Isvara/Shiva)
- Phra Ram (Rama) son of king Tosarot(Dhasarath) of Ayutthaya and Incarnation of Pra Narai
- Nang Sida (Sita) wife of Phra Ram, embodies purity and fidelity. Incarnation of Nang Lakshmi
- Phra Lak(Lakshman) Pra Phrot (Bharata) & Pra Satrut (Shatrughna) half-brothers of Pra Ram
- Hanuman—God-king of the apes, who supported Pra Ram and acted as the APE GENERAL
- Thotsakan (Ravana, from *dashakantha*)—King of the Demons of Lanka
- Intharachit (Indrajit)—A son of Thotsakan
- Phiphek (Vibhishana)—enstranged brother of Thotsakan
- Pali Thirat (Bali)—King of Kitkin, elder brother of Sukreep(Sugriva) and uncle of Hanuman
- Ongkot (Angada)—Ape-prince of the Pali Thirat and Nang Montho, cousin of Hanuman

<<<<<<< §µΨΦ∞Жλπ∐δ >>>>>>>

None of the Scriptures apart from Hinduism, Olmec and Mayan civilization mentioned about air crafts or space ships used which I used to wonder, the marvel of engineering and technology used in those days to fly a aircraft, the spaceship carried the humans from one planet to another planet in and the aliens visiting the earth depicted in Hollywood films which they named it as UFO, was exactly mentioned in indian Scriptures the lords of different lokas(worlds) visiting the earth in VEDAS.

<<<<<<< §µΨΦ∞Жλπ∐δ >>>>>>>

**It is ancient belief in Hinduism, Sometimes the modern day UFO citings or the spaceships seen may have doubted to come from Talatala-loka is one of the svarga-loka of different planetary system on vedic decodings. Talatalaloka is 11th Loka above 10th is the Sutala loka and below is Mahatala-Loka 12 th loka ours is Bhurloka is 6th. Talatala Loka is considered of beings with great material knowledge.**

<<<<<<< §µΨΦ∞℀λπℿδ >>>>>>>

In the midst of all the above understandings we certainly need to overlook how history is repeating itself in different forms.

Ever since man started inventions, later started replacing earlier, Human mind started thinking and thoughts got wiser and smarter, which seemed to had the sense of showing off their primitive group is greater than others and led to conquering the other groups as their slaves. Violence unleashed and battles fought and many killed and looted the belongings and it continued until man decided to stop eating man (cannibalism) and decided to farm the lands and grow cattles and poultries started searching for alternative food and turned his place as vegetated land.

Civilization grew again and the history continued to repeat itself only this time from Primitive behavior to civilised world, In civilised World, it was kings who continued the same patterns of behavior and out fought other civilizations and allowed defeated kings to pay homage and taxes and required to lend their army at the time of battles or war.

In Later Era when kings lost their kingdoms and during the transition period approximately from 13th century to early 19th Century the Kings and empire's Kins turned their Dynasties to Monarchies, occupying the land part and allowed the people to live in their land by paying homages certainly meant the leasing of land for agriculture and farming (Colonisation).

It was in the early 19th Century Monarchies and Dynasties decided to border their land area and named after which is registered for the further conflict on land occupying for that matter of Fact the British Monarchy Continuing to till date holding the British colonies and Union Territories spilled over in the midst of modern Republican and Democratic world of countries.

While observing, it is understood whether it is stone age primitive people or the modern day world the concept of war is never changed and the concept of war now turned in the name of War of Religion and Terrorism, the same thoughts happened to remain with primitive people then and now to the present governments, this time the underlying factor is the money, growing population and shortages. The present day world is now enjoying the economic boom, technology breakthrough in Agriculture, More rainfall and floodings along with extending deserts and river dryups and famine and poverty all happening simultaneously in different parts of the world trying to balance the ecology in the given environment.

Going thru all the changes the Myth remained in every religion and people following to that religion still remains on high note, is it good or bad really? We really need to know what is the meaning of Myth or Mythology—It's a belief worshipped of an imagined Person/Situtation/Institution when exaggerated or fictitious. Even India having a rich Vedic Science is no exception when it comes to mythological Beliefs. Mythological beliefs are often believed in fear psychology led to trauma shock treatements developed over the years.

Many cultures to have worshipped often on Demon Gods, Dark magics and other frightening ritual practices. To my thoughts all these ideas developed during primitive and medieval period to keep the community in control, discipline and fear which paved way to avoid unnecessary fights or hatredness or politics in and within the Groups or Community.

Later on the success of the practices which brought in great fear and discipline the anticipated results of the wise devised it was acheived, Many Civilisations had temple and monuments constructed with weird and scary shapes and sacrificial rituals performed till date in almost all parts and corners of the world. On many situations the belief in this mythical demon gods gained momentum and become the part of the religion today. Any so called Famous religions in the world today have a separate conceptuals of SATAN God.

The word **Voodoo**, which has many different names and spellings (like *Vodun, Vodou, Voudou, Vudu, Vodoun*) is the name of a West African religion. Voodoo is animist and spiritist, and a lot of Voodoo beliefs have to do with ancestor spirits. Some of it is based on Catholic beliefs. In Voodoo many gods and spirits are prayed to or called on. Both spirits of nature and of dead people are important. The spirits of family member who have died are especially important. Voodoo often has rituals with music and dancing. Drums are used to make most of this music. In Voodoo people often believe that a spirit is in their body and controlling the body. Having a spirit come into is wanted, and important. This spirit can speak for the gods or dead people you love, and can also help to heal or do magic.

<<<<<<< §µΨΦ∞ℋλπЏδ >>>>>>>

India is the land of **confused worship only when it comes to worship of Demigods or Satan gods**, Aghori Sadhus at kasi, India, found to be eating the half burned dead bodies, in which they have their own cult and rituals as they say the bodies these aghori sadhus eat will get liberated from next birth or reincarnations, while i tried understanding their cult what I witnessed is worshipping of Godess KALI before performing this cult of eating dead bodies, as mythical stories depict Godess KALI as the godess who took her will and purpose to kill RAKHTABIJA identified to be an ASURA (evil minded warrior) to save the innocent people from him, In the due course when battle got intense, people around realized that Godess KALI was killing and in the process, drinking blood of the defeated, gets carried away from regular human emotions, and she simply started slaughtering everything and anything got on her way and she became a monster from being a Godess, in observing the situation, people approached LORD SHIVA and SHIVA deviced a plan to arrest the on-going chaos, attempted the plan by pretending to be dead underneath Godess KALI's feat, made her understood the state of her mind and calmed Godess KALI down as per myth. Many ritual practices followed by Indian women in connecting to this episode worship Kali believing the godess will give ladies the power of ferociousness and courage and stand against evil thoughts and acts,

What is happening in contrary, in many states of india especially southern part of india, this ritual idealogy was misconceived along the way due to misinterpretations, seemingly few women and ladies who believe in deep myths of Godess KALI found to be becoming Hysterical paranoid and schrizophenic paranoid raising mental and nervous problems and straining the family members by spoiling the peaceful life, leaves me thinking what in earth these kind of mythical beliefs rooted into religion spoiling the middleclass Indian family life, Genuine cases, to be immediately treated at hospital and may require even psychological treatments.

There are some other funnier cases also witnessed, that is unique way of indian women purposely acting paranoid and weird as their objective is to establish the control and fear in their own house in process, establishing domination on household activities and even with husband and relatives.

"I laughed out loud". I ask now what difference between the cases of demonic possession and Mythical god possesion, both were almost in the same state of mind, It's the belief system causing the unnecessary headache and nothing else leaving the highly conceptual ideologies of Hinduism to the critics tongue.

<<<<<<< §µΨΦ∞ЖλπЏδ >>>>>>>

Lord SHIVA the iconic god understood to be in the human form of RUDRA. However one should remember Rudras were 11 in number as in Hinduism. **KAPALI, SHIVA, PINGALA, BHIMA, VIRUPAKSHA, VILOHITHA,**

**SHAHSTHA, AJAEKAPADA, SAMBHU, CHANDA and BHAVA.** But from original Script decoded from Vedic Samhihatas, **explains Rudras scientifically identified with the forms of forces and energy, RUDRA is with forms of STORM, HEAT, LIGHT, FIRE, RAIN, AIR, GASES, FERTILITY, SOIL, VOLCANICs and EARTHQUAKES** reason each one being the properties of NATURE, mentioned above if gets slightly disturbed from its level of balances may feared to have sequencing destructions. People feared RUDRA as it may bring massive scale of devastation in no time, worshipped shiva as the human form of RUDRA, as human beings, the primitive people witnessed him as the power of destruction as he is very skillful in hunting tactics, wartime master stratagiest and had great weapons development skills and mass destruction techniques, Civilised Kings took his help to defeat the evil practices uprising among the people of other semi civilized and Barbarian kings. The belief system and sentiments among the human beings have gained to their stature of stubbornness, inflated ego in an individual had level of perception which resulted in narrow mindedness.

Perhaps it makes me to think sometimes is this why VEDAS gave us the scientific knowledge along with the blend of superstitious belief in the form of UPANISHADS, PURNAS, ITIHASAS to keep the underlying scientific facts hidden underneath the concepts of stories passed on to the generations in Hindu religion. For the noble cause it brings in discipline and fear having mythical images, beliefs seems to be harmless, but if it turns out to be business, fraudulents, deceiving others etc., then it is highly punishable even in Indian courts of law.

In contrary, apart from Mythical Satan Gods, the angels or brahmarishis from other parts of world mentioned in many religions also seen in parallel idealogy

| SAPTARISHIS (HINDUISM) | ARCHANGELS (JUDAISM) | ARCHANGELS (CHISRITIANITY) | ARCHANGELS (ISLAM) |
|---|---|---|---|
| KRATU | GABRIEL | GABRIEL | JIBRAIL |
| PULAHA | MICHAEL | MICHAEL | MIKAIL |
| PULATSYA | RAPHAEL | RAPHAEL | ISRAFIL |
| ATRI | URIEL | URIEL | MUNKAR |
| ANGIRAS | RAGUEL | SIMUEL | NAKIR |
| VASISHTA | REMIEL | ORIPHIEL | RIZWAN |
| BHRIGU | SARAQUIEL | RAGUEL | KIRAAMAN/ KATIBIN |

In Similar, from other parts of world were also mentioned in various religions (the world of the Gods)

| 7 Heavens/Universe HINDUISM | 7 Heavens/Universe CHRISTIANITY | 7 Heavens/Universe ISLAM |
|---|---|---|
| VAIKUNTHA Ultimate heaven of no return SATYALOKA | VILON | RAFI' |
| TAPALOKA | RACKIA | QAYDUM |
| JANALOKA | SHECHAKIM | MARUM |
| MAHARLOKA | ZEVUL | ARFALUN |
| SWARGALOKA | MAON | HAY'OUN |
| BHUVARLOKA | MACHON | AROUS |
| BHULOKA | ARAVOT | AJMA' |

The above informations shall be lending the discerning insights about our world and the changes in the way of living, over milleniums also lighting us for clear visions to our mind how the mythology developed and people became god abiding nature and how religion grew and what religion taught us and where we are proceeding, holding the religion in our hearts and it is time to refresh our thoughts and proceed further to next chapter to understand what we need to guide the younger generations properly and safeguard them falling prey to few fundamentalists preaching the religion with wrong intentions and promoting the terrorism. Anybody should be able to know, how and what to choose from the religion based political parties and government of a nation to strive and succeed further economically as well as literacy based nation, I mean here is not that a man should be a literate by learning how to read and write, literacy i mean here that a man should be aware of the world around him and to develop his persipicacious while being ardent about what the world is impelling today.

# 5

# *Taking time for Critical thinking*

While comprehending the basics I always had thoughts why archeaologists and geologists sometimes use to time the historic events wrongly and when the literary works dates available clearly when the Rama avatar happened in india in treat yuga which is atleast before 1.7 million years approximately why these geologists couldn't give not more than 5000 years BC and with great inaccuracy Indus valley civilization was dated only 3500 years BC, ok understood these geologists have limited tools and limited access for the interpretation of actual scriptures in different languages and holding the yard stick of time in BC (Before Christ), So they take the calendar system from the birth year of Jesus Christ, I really don't understand what is the preconceived notion these geologists and archeologists have to time it from the birth year of jesus, because they follow Gregorian calendar system as they consider it is the first day of jesus Christ born is world birth date or the new era in the world began,

Loud and clear to geologists and archeologists there were many era have passed before Jesus Christ, there were fantastic dynasties and civilizations lived before the Anno Domini(AD) era, civilizations and dynasties survived millenniums before the life time Christ. Technically observing and terming it was the era of Jesus Christ's popularity and his miracles performed and considered as messenger in his days was the most short lived in the entire history.

## Hurrian Religion (Old Hinduism) in North America:

It is in Grand Canyon is a steep-sided canyon carved by the Colorado River in the United States in the state of Arizona. The Grand Canyon is 277 miles (446 km) long, up to 18 miles (29 km) wide and attains a depth of over a mile (6,000 feet or 1,800 meters).

Vishnu Basement Rocks is for all Early Proterozoic crystalline rocks (metamorphic and igneous) in the Grand Canyon region. Early Proterozoic—1.680 to 1840 billion years (age used by National Park Service, Mathis and Bowman, 2005) **Vishnu Schist**—Quartz-mica schist, pelitic schist, and meta-arenites of metamorphosed, arc-basin, submarine sedimentary rocks. About 1.75 billion years old.

**Brahma Schist**—Consists of amphibolite, hornblende-biotite-plagioclase schist, biotite—plagioclase schist, orthoamphibole-bearing schist and gneiss, and metamorphosed sulfide deposits. Mafic to intermediate-composition metavolcanic rocks. About 1.75 billion years old. Both adding to the surprise that the inhabitants even moved from India to North America established and survived as religion.

<<<<<<< §µΨΦ∞ЖλπЏδ >>>>>>>

**Who did exegesis it wrongly? Who were the greedy minds making money by hiding facts of original scriptures?**

In the gospels today with many versions, still continuing with missing years in the life of jesus almost 14-15 years of life time is missing in scriptures (gospels) in all four early gospels of John, Matthew, Luke and Peter. If it was not intentionally covered up, if they genuinely know that jesus was really missing in neighbourhood and doesn't know really what happened to him during those missing years, Iam sure if he would have told to his people where he gone for some 14 years when he came back is sure, or is it intentionally hidden while writing the gospel or it is hidden during the revised editions.

Surprising details again was not explained in detail in the episode of "The Gift of magi/Biblical Magi", and the description were almost same in the other 3 gospels of John, Luke and Peter. Yes we all remember the opening scene in the Hollywood movie "Ben-Hur" released in 1959.

**Matthew 2:1-2:**

"Now when Jesus was born in Bethlehem of Judaea in the days of Herod the king, behold, there came three wise men from the east to Jerusalem,

Saying, Where is he that is born King of the Jews? for we have seen his star in the east, and are come to worship him."

**However, Matthew 2:11 states that they visited him at his house:**

"And when they were come into the house, they saw the young child with Mary his mother, and fell down, and worshipped him . . ."

**Matthew 2:16 states that Herod had all the children two years and under slain, according to the time he learned from the wise men.**

"Then Herod, when he saw that he was mocked of the wise men, was exceeding wroth, and sent forth, and slew all the children that were in Bethlehem, and in all the coasts thereof, from two years old and under, according to the time which he had diligently inquired of the wise men."

The Christmas season traditionally ends on Epiphany, January 6, the 12th day after Christmas. In Western Christian churches, this day known as Twelfth Day or Three Kings Day celebrates the visit of the wise men to the young Jesus.

**But who are these 3 wise men really from far east—far east is it referring to asian region?** followed a star, followed a moving star? How can somebody follow a moving star in space to its speed in earth distance I mean is the distance travelled in so much speed? or **is it really pointing to Astrological science or Jyotish shastra was considered as greatest science practiced in asian region.**

However mother mary should be knowing who are those 3 wisemen? or atleast the neighbours would know as they witnessed the visit of 3 wise men should have helped with the details of those 3 wise men like what are their names? How do they look like? which part of the world they are from?

We all know it was King Herod sent these 3 wise men, but these 3 wise men were from far east? but king herod from south of Judea (Palastine)? Did King herod hired these 3 wise men to identify the birth location? Why did King Herod hired wised men from asia not from anywhere else??

If these missing details would have been explained elaborately Christianity would have been in a very different dimension today as a religion. The conspiracies is not in Christianity alone about missing periods or hidden facts as it is stretching way back in Hinduism as well.

<<<<<<< §µΨΦ∞Жλπμδ >>>>>>>

As we all know that the very similar kind of conspiracy in Hinduism when it comes to Shaivism in the missing episodes or hidden episodes. We all know after the killing of demon king Tharakasura, it was in scriptures that Shiva and Parvati along with their son lord ganesh lived happily ever after on Mt.Kailash.

But the scripture does not continues with episodes of the end or death of Shiva and Parvati, or there is no further detailed explanation of the family linage is not specified, what happened to Lord Ganesh son of Shiva, It was there in scripture that karthikeya first son of Shiva departs from his family due to fued with his brother ganesh on an intellectual test in which karthikeya gets defeated, angry brother leaves his family and gets settled in southern state in India at Tamilnadu where he gets married to two women Valli and Devasena, Again Karthikeya's legend and linage gets discontinued, abruptly like Shiva and Parvati legend.

Why when a family is being worshipped almost like god?!!! the events happened have carried the legend all the way to till day from treata-yuga (millions of years before), and have so many temples built to the Shiva and his family members and many millions of worshippers across the world, never thought of this abrupt discontinuance? or Nobody bothered to question the discontinuance? or not willing to question about it? or the worshippers of this cult is so much convinced to worship the LORD SHIVA and his family inside a temple without questioning discontinuance in legend's history of lineage? These Demi gods have had their images and Idols lifted to the level of divinity where so many temples raised all over india have hidden episodes??.

Are the preachers of this cult did not even worried about telling the followers and worshippers to let know about the continued history of this family legend?

If it is this cult gained popularity in so many years? As an author of this book let me be the first person to question the preachers of this cult as an enquiry of the worshipper to know about the discontinuance of the history?

<<<<<<< §µΨΦ∞ЖλπЏδ >>>>>>>

Very similar doubts in Islam?

Muhammed the prophet gets married to very wealthy widow named Khadīja al-Kubra considered the mother of islam. Khadija's mother, a member of the clan of the Quraysh and a distant relative of Muhammad. Khadija married three times and had children from all her marriages. To her first husband she had two sons. Upon all the marriages Khadija was left as widow.

Khadija became a very successful merchant over time. It is said that when the Quraysh's trade caravans gathered to board the vehicle upon their trade journey to neighbouring countries, Khadija's number of caravans were more than the caravans of all other traders of the Quraysh put together. Khadija looking for the business manager as the trade figures grew, hired Muhammad, was then a young man also younger than khadija. Khadija was impressed by honest nature of Muhammad and the skills he displayed in operating business, with the result that he brought back twice as much profit as Khadija had expected.

Maysara an appointed assistant to Muhammad, upon returning from trade journey informed Khadija as Muhammad was tired and while sleeping under a tree, he had seen two angels standing above Muhammad creating a cloud to protect him from the heat and rays of the sun. Khadija consulting **Waraqah ibn Nawfal** and waraqah said that if what Maysara had seen was true, then Muhammad was in fact the prophet of the people who was already expected. Khadija entrusted a friend named Nafisa to approach Muhammad and ask if he would consider marrying Khadija, a widow after 3rd marriage.

Being the events described in the scripture, on citing the reference, Muhammad and Khadija were married successfully for twenty-five years. It can be speculated that this was because khadijah was of a higher social status than Muhammad and could therefore demand fidelity from him. Khadija goes and enquires about the incident of Angel Gabriel appearance, to Waraqah ibn Nawfal? why to Waraqah/ who is Waraqah? Waraqah ibn Nawfal was the paternal first cousin of Khadija and the first wife of prophet Muhammad. According to the Islamic sources Waraqah was a priest and one who had made studies of the gospels and the old testament scriptures. Waraqah frequently prayed at Kaaba and began to read Bible and also learned Hebrew language. Muhammad growing wiser by experience as a man, Waraqah's knowledge of the sacred scriptures developed as well. Years later, when told of Muhammad's first revelation

(Sura 96: 1) Recite in the name of your Lord who created . . . .
(Sura 96: 2) Created man from a clinging substance . . . .
(Sura 96: 3) Recite, and your Lord is the most generous . . . .
(Sura 96: 4) Who taught by the pen . . . .
(Sura 96: 5) Taught man that which he knew not . . . .

Waraqah recognized his call to prophecy as authentic, its Waraqah acknowledging the knowledge in scriptures what prophet the Muhammad had possessed the knowledge about prophecy, Waraqah saying "Reached to him the greatest Law (mentioning the Gabriel) that came to Moses, surely he is the prophet of this people.

So Waraqah alone is enough to validate muhamad as prophet? Nobody really questioned about waraqah's validation?

As everybody surprised Waraqah, upon accepting Muhammad's prophecy, remained a Christian and, in later accounts, was counted among Muhammad's companions.

Muhammad is later said to have said of Waraqah "Do not slander Waraqah ibn Nawfal, for I have seen that she will have one or two gardens in Paradise?

What in earth he says "Gardens in Paradise"? which means he himself believed that he is more than a messenger and equal to god, where he has the liberty to provide 1 or 2 gardens in paradise? or is he mentioning the land he won after war? it confirms that he still had love for his first wife Waraqah. or is it a praise or a kind of compliment for her for having acknowledged that jibreel's (Gabriel) visit and allowing muhamad for marrying the wealthy women khadija?

**The entire concept seemingly to be a family drama between muhamad, khadija, Waraqah and maysara, except maysara, understandably the whole drama is revolving around 3 family relatives?**

**Is khadija and muhamad marriage is the marriage of convenience for business interest to safeguard the business, Khadija clearly capitalizing muhamad selling and business operational skills? and Muhamad clearly capitalizing the wealth of Khadija by marrying her being elder in age and widow after third marriage.**

Did Jibreel (Gabriel) instructed muhammad to preach and propogate the islam religion as per scripture? or is it the idea Muhamad derived out, based on confusion and to stay away from the Judaism and Christianity due to tiredness of the conspiracies that two different ideologies approached in life? or muhamad was rich enough to convice the people so, that he is the messenger of allah confirmed by Jibreel (Gabriel) with only one eyewitness being the paid employee? why jibreel (Gabriel) did not opted to show himself when he was in public place? gives the impression of complete drama is happening in isolation? and isolated locations? if jibreel was to make revelation he could have revealed it in a church or juda temple or somewhere where many people present? ok **it is clearly a Business-family drama**? Great it is seemingly that muhamad in sura on heavens does not conspire with Judaism or Christianity? it is clear he wanted to be worshipped as demi god and wanted a separate group of followers and worshippers?

But why? To build a business empire? or have their own network of assistants and franchisees for his business empire? or to conquer the massive part of land as a separate country to have own religion and governance? in the process of doing it, there is war in between occurs to establish the ownership of governance? it appears to be all in one hit by propogating himself as messenger from God?! Good god Allah?!!!

Very same behavior seen in Islamic dictators in Modern day, self announced kings and dictators at one point of time naturally looking for war on weaker neighbouring governance or country is highly questionable how far they have entered and understood the world of republic and democratic countries in 21ˢᵗ century?

It appears self announced dictators were clearly not giving their country governance to public to decide and not willing any outside interference?

In contrary their willingness to have muti-national corporations to do business under their command and stronghold and support their style of governance?!!???.

Unlocking the grids, from the frames these selfish religious fundamentalist group, Greedy Money Makers, self announced dictators, self acclaimed presidents setting the traps by using the unique tool of false interpretations and hiding the facts of scriptures in religion and diverting people away from the facts.

<<<<<<< §μΨΦ∞Жλπμδ >>>>>>>

**LETS also explore what we have in bags offered by Modern Science.**

One case goes back to June 3, 1965, during the Gemini IV mission when Ed White performed the first space walk by an American. At one point during the flight, command pilot Jim McDivitt, pictured at right, spotted an unusual object in space.

In his words, "At the time that I saw it, I said there was something out in front of me or outside the spacecraft that I couldn't identify, and I never have been able to identify it, and I don't think anybody ever will," McDivitt said back in 1975.

"We were drifting flight and my partner, Ed White, was asleep. I couldn't see anything out in front of me except just the black sky. And as it was rotating around, I noticed something out in front that was a white cylindrical shape with a white pole sticking out of one corner of it—it looked like a beer can with a smooth pencil sticking out." What a funny decription by an astronaut??

In another case presented on "NASA's Unexplained Files," the crew of the space shuttle Columbia attempted to deploy a satellite attached to a 12-mile-long cable or tether during their 1996 mission, STS-75. The tether is miles away and is 12 miles long, and these objects appear to be miles across, and this is what excited the UFO community so much—this was their smoking gun."

Unconvinced, Dantonio created an experiment where he recreated the tether, hung small particles in front of it and filmed them. "The particles are brightly lit and are also out of focus, making them appear very large," Dantonio said. The resulting optical illusion shows the particles seeming to pass behind the makeshift tether, even though they're actually passing in front of it. Great however modern science fail to confirm even the details clearly what they have seen actually or reported.

Coming back to Indian puranas, we are speaking proudly about VIMANAS in RAMAYANA, called flying chariots of King Ravana, have been scripted, King Ravana used to ride vimana across oceans and seas visiting India, It is doubtful the ancient space ship technology is stolen from ancient treat yuga civilizations as per veda or UFO beings from space visitors of other planet have given us the technology of flight systems and aerodynamics shall be debated really or some body left the earth and upon returning, the calculation of space time missed completely to see earth changed in centuries where all the dynasties and kingdoms fell and communications got disconnected??.

<<<<<<< §µΨΦ∞ℑℵλπЦδ >>>>>>>

Vimanika shastra by maharishi Bharadwaaja is scripted in Vedas with diagrams, plans and measurements is carefully studied

"The ancient Hindus could navigate the air, and not only navigate it but fight battles in it like so many war-eagles, combating for the domination of the clouds. To be so perfect in aeronautics they must have known all the arts and sciences relating to the science, including the strata and currents of the atmosphere, the relative temperature, humidity, density and specific gravity of the various gases . . . ."

Col. Olcott in his lecture in Allahabad in 1881.

The Rig Veda, the oldest document of the human race includes references to the following modes of transportation,

**(Rig Veda 6.58.3);** Jalayan—a vehicle designed to operate in air and water a vehicle that operates on ground and in water.

**(Rig Veda 9.14.1);** Tritala—a vehicle consisting of three stories.

**(Rig Veda 3.14.1);** Trichakra Ratha—a three-wheeled vehicle designed to operate in the air.

**(Rig Veda 4.36.1);** Vaayu Ratha—a gas or wind-powered chariot.

**(Rig Veda 5.41.6);** Vidyut Ratha—a vehicle that operates on power.

Ancient Sanskrit literature of vimanika shastra is full of descriptions of flying machines—Vimanas, found from many documents is evident that the Brahma rishi Agastya and Bharadwaja had developed of aircraft construction.

The "Agastya Samhita" gives us Agastya's descriptions of two types of aeroplanes. The first is a "chatra" (balloon) is to be filled with hydrogen. The process of extracting hydrogen from water is described in elaborately and the use of electricity in achieving this is clearly described. This description appears to be a primitive type of plane, useful only for escaping from a fort when the enemy had set fire to the jungle all around. The name used as "Agni-yana". The second type of aircraft described is appears to be of parachute. It could be opened and shut by operating chords. This aircraft (parachute) has been described as "vimanad-vigunam".

Aeronautics or Vimanika Shastra is a part of Yantra Sarvas(v)a of Bharadwaja. Vimanika shastra deals about aeronautics, including the design of aircraft, the way they can be used for transportation and other applications, in detail. The knowledge of aeronautics is described in Sanskrit more than 80 sections, 6 chapters, 400 principles and more than 2700 slokas. Great sage Bharadwaja explained the construction of aircraft and scripted shastra how to fly it in air, on land, on water and use the same aircraft like a sub-marine. He also described the construction of war planes and fighter aircraft to carry weapons and explosives.

Vimanika Shastra explains the metals and alloys and other required material, which can be make an aircraft non-perishable in any climatic condition. Planes which will not break (abhed-ya), or catch fire (adaah-ya) and which cannot be cut (ached-ya) have been described. Along with the treatise there are diagrams of three types of aeroplanes—, "Shakuna", "Tripura" and "Rukma" vimanas

The aircraft is classified into three types—Mantrika, Tantrika and Kritaka, to suit different yugas or eras (ages of time). In Satyuga, it is said, dharma (righteousness) was well established. The people of this yuga (period of time) had the divine powers to reach willing places using only their Ashtasiddhis (eight super-normal powers).

The aircraft used in Tretayuga are called Mantrika-vimana, flew by the power of hymns (mantras). More than 20 varieties of aircraft including Pushpaka Vimana belong to this era.

The aircraft used in Dwaparayuga were called Tantrikavimana, flew by the power of tantras. More than 50 varieties of aircraft including Bhairava and Nandaka belong to this era. The aircraft used in Kaliyuga, the on-going yuga, are called Kritakavimana, shall fly by the power of engines is what mentioned in the Vimanika shastra scriptures.

Bharadwaja states that there are more than 25 secrets protected or scripted in the aeronautical science. Some are amazing and some secrets indicated an advance level of technology even beyond our own times. For instance the secret of "para-shabda graaha", a state of art designed room for decoding the communication datas or systems, has been explained by an electrically worked sound-receiver that did the communication interpretable. Manufacturing different types of instruments and putting them together to form an aircraft are also described.

Evidences for aircrafts existence are also found in Arthasastra of Kautilya (3rd century B.C.). Kautilya mentions amongst various businessmen and engineers the Saubhikas, as pilots operating aircrafts in the sky. Saubhika is the name of aerial flying city of King Harishchandra and the term Saubika means one who knows the science of flying an aerial city. Kautilya uses another important term Akasa Yodhinah, which actually meant as pilots who are trained to attack from the sky. Aerial chariots, in whatever form it might be, was so well-known that it found a place among the royal edicts of the Emperor Asoka also operated the aircraft or aerial chariots during his reign from 256-237 B.C.

**The Academy of Sanskrit Research in Melkote, near Mandya, was commissioned by the Aeronautical Research Development Board, New Delhi, to take up a one-year study, Non-conventional approach to Aeronautics, on the basis of Vaimanika Shastra. As a result of the research, a glass-like material which cannot be detected by radar has been developed by Prof Dongre, a research scholar of Benaras Hindu University. A plane coated with this unique material cannot be detected using radar.**

<<<<<<< §µΨΦ∞Ж𝜆𝜋Ц𝛿 >>>>>>>

The Indian science of aeronautics and Bharadwaja`s research in the field of Aerodynamics was successfully made for test-fly by an indian over hundred years ago. In 1895, atleast 8 years before the Wright Brothers's first flight at North Carolina, Shivkar Bapuji Talpade gave a demonstration flight on the Chowpati beach in Mumbai.

With this plane this pioneer airman of modern India gave a demonstration flight on the Chowpatty Beach in Mumbai in the year 1895. The machine attained a height of more than 1100 feet and landed safely. This event was witnessed by Sayaji Rao Gaekwad, the Maharaja of Baroda and Justice Govind Ranade and was reported in The Kesari a leading Marathi daily newspaper. It is reported in the newspaper that the witnesses were impressed by the invention and appreciated and acknowledged the talented inventor.

This success of an Indian scientist was not at all supported by the Imperial rulers. The British Government warned, and Maharaja of Baroda withdrew his support of helping Talpade. It is said that the remains of the Marutsakthi (wind energy technique or the possible propulsion technique) were forcefully sold to abroad foreign parties by relatives of Talpade. But his efforts is known for the greatness of Vedic Shastras was recognised by Indian scholars, gave **Talpade the title of "Vidya Prakash Pra-deep"**. <u>**Talpade died in 1916 unhonoured in his own country**</u>.

Instead Wright Brothers were honored for their achievements, we should think of Talpade utilised the ancient knowledge of vimanika shastra, to fly an aircraft, eight years before Wright Brothers.

<<<<<<< §µΨΦ∞ЖλπЦδ >>>>>>>

A study conducted by aeronautical and mechanical engineering students at Indian Institute of Science, Bangalore in 1974 concluded The Rukma Vimana was the only one which resulted in possibility of flying. It had long vertical ducts with fans on the top to suck air from the top to bottom, generating a lift in the process.

Its existence was first announced publicly in a 1952 press release by G.R. Josyer, who had founded his "International Academy of Sanskrit Research" in Mysore the year before. In the foreword to the 1973 publication that contained the full Sanskrit text with English translation, Josyer quotes a 1952 press release of his which was **"published in all the leading dailies of India, and was taken up by <u>Reuters and other World Press News Services</u>"**

In 1991, the English portion and the illustrations from the Josyer book were reprinted in *Vimana Aircraft of Ancient India & Atlantis* as part of the *Lost Science Series in which* 8 chapters treat the following:

1) The process of constructing aeroplanes, with fire resistance.
2) The technology of making planes invisible in air while flying.
3) The technology of hearing conversations and other sounds in enemy places (the radio signal receiver)
4) The technology of ascertaining the direction of enemy planes approach.
5) The technology of making persons in enemy planes lose consciousness.
6) The technology of destroying enemy planes.

The propulsion of the Vimanas is "Mercury whirling fluid engines", apparently a concept similar to electric propulsion. The evidences for this "mercury whirling fluid engine" found in the *Vimanika shastras*, script of aerodynamics.

If Vedas is predating all technologies and highlighting the presence of ancient technologies, in its scriptures is the oldest available scripture in the world. How precise and specific it should be in the field of COSMOLOGY, Undoubtedly the scientists in the world today cannot bypass the theoretical metaphysics where in Indian Vedas steps in to shed enormous light for the practical understandings, However to step further in next chapter, i will be dealing more on cosmology, the GALAXY, BIG BANG, SUPERNOVA and much much more coming up . . . .

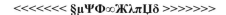

<<<<<<< §µΨΦ∞ЖλπЏδ >>>>>>>

# 6

# *The Creation and its Evolution in Vedas and Modern science*

Vedic explanation of Big Bang theory in modern day as we know it as Collision of Galaxies due to gravitational pulling, However the philosophical idea of creation seemingly the closest to Modern science understanding in the Vedas. Creation of universe was discussed by Vedic rishis since long time. Rigveda Vol X. clearly refers to the creation of the Universe.

<<<<<<< §µΨΦ∞ЖλπЏδ >>>>>>>

**Rig veda: X.129.1: (actual Sanskrit hymns spelled in English Language)**

*Nasad-asinno sadasit(t)h adanim nasidhh rajo no vyoma paro yat |*
*kimavarivah kuha kasya sharman(n) ambhah kimasid gahanam gabhiram ||*

**Persipicacious Meaning:** It was neither existed or non-existed. There was neither the realm of space nor the sky which is beyond. What stimulated? Where the stimulation occurred? In whose permission? Was there ocean, deep without bedrock? and it was there always . . . .

**Rig veda: X.129.2: (actual Sanskrit hymns spelled in English Language)**

*Namrth(yur)asidamrtham na tarhi na rathr(y)a ahna asith prakethah |*
*anidavatam svadhaya tadekam tasmaddhananna parah kim chanasa ||*

**Persipicacious Meaning:** There is either death or immortality then. There was no differentiating indication of day or night. But there is something breathed, without wind, on its own pulse. There is nothing beyond and it is not explainable, and it was there always . . . .

**Rig veda: X.129.3: (actual Sanskrit hymns spelled in English Language)**

*tama asi(t)thamasa gur magre praketham salilam sarvama idam |*
*tuchc(h)hyenabhvapihitam yadasithtapas asthan mahina jayataika||*

**Persipicacious Meaning:** In the beginning there is nothing but Darkness, With no differentiating sign, all this was like ocean (referring to milky way). The force of life is filled with emptiness, It all happened due to the power of heat occurred due to gravitational pull . . . .

**Rig veda: X.129.4: (actual Sanskrit hymns spelled in English Language)**

*Kamastad agre samavartat adhi manaso rethah prathamam yadasith |*

**Persipicacious Meaning:** Desire arise and it is the beginning, It was the first seed of mind.(Spark of thought and the process of thought leading to desire).

**Images of galaxies getting closer for collision due to gravitational pull**

**Images of massive collision of galaxies of intense gravitation**

<<<<<<< §µΨΦ∞Жλπυδ >>>>>>>

**As Per Vedic Thoughts to Modern science**

**Gravitational pull as <u>primary source (Narayana)</u> of two existing Galaxies**

**Occurrence of Collision and after Big Bang creating Cosmic Space, time and Huge Black Holes (Brahma), in due course of process of creation, later a strong beam of light and heat created (Shiva and Shakti) creating the artificial magnetic field holding the neighbouring planetary system intact forming an orbit.**

<<<<<<< §µΨΦ∞Жλπυδ >>>>>>>

**Lets see what fervent the modern science have for us**

While the Andromeda Galaxy contains about 1 trillion stars and the Milky Way contains about 300 billion stars, the chance of even two stars colliding is negligible because of the huge distances between the stars. For example, the nearest stars to the Sun is Alpha Centauri A, Alpha Centauri B and Proxima Centauri, (these names given by modern day astrophysicists to be remembered and these names were not used in Vedas) about 4.2 light-years solar diameters away. No Surprise one of the Centauri is the planet exploration which was shown in the famous Hollywood movie AVATAR.

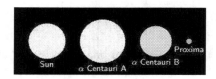

The Milky Way and Andromeda galaxies each contain a central supermassive black hole, these being Sagittarius A. These black holes will converge near the center of the newly formed galaxy, transferring orbital energy to stars that will be moved to higher orbits by gravitationally interacting with them, in a process that may take millions of years. When they come within one light year of one another, they will emit gravity waves that will radiate further orbital energy until they merge completely.

The Reason being Hinduism is considered as the most scientific religion in the world is due to the facts and the strong intellectual calculations done in the field of Astrology. The movement of stars and its changing celestial constellation and the movements interjunting into the orbit of other planetary systems or solar systems which was well conceived in Indian astrology called JYOTISH VIDYA or JYOTISH SHASTRA. As the author of this book i take my stands strongly westerners didn't understood the cosmology in depth as Indian Jyotish rishis understood and lived, gave us the fantastic Jyotish shastra what we believe and follow it in india even today,

<<<<<<< §µΨΦ∞ЖλπЏδ >>>>>>>

The distance between the Galaxies and the collision and after collision effects have been clearly mentioned in Vishnu Purana, about the timelines is being explained in Vedas as (MANVANTARA).

So What is MANVANTARA? A Period of Cosmic Time, According to Hindu mythology there is 71 MAHAYUGAS in 1 MANVANTARA constituting 306,720,000 years the world is destroyed and recreated and the recreations may occur in the same planet or the design of creation shall get evolved in an another planet where the climate and environment is suitable for life creations. The 14

Manvantras is being assigned with a name of MANU is actually a time scale, but for better understanding these were given the human forms and depicted as a story. Different Sets of GODS and Rishis (different set of star constellations) are born in successive manvantaras including 7 SAPTA RISHIS. Gods and Rishis to be born in human form on the important occurrences in the world happening to stabilizes the world from getting demolished by evil, selfish minded human or group of humans were always remains as stars in the sky and watching the events in earth was believed in Hindu Mythology. Many religion and early civilizations believed in this concept of Idealogy and always worshipped the sky believing god will be incarnated in Human form to save the world, lets say it is the standard belief system all around the world

<<<<<<<< §µΨΦ∞Ж λπЦδ >>>>>>>

The following table will make you understand the depth of Time scales what we are dealing with.

| Manvantara | Manu | Incarnation of Vishnu as | Tribes/Demi Gods | Rishis |
|---|---|---|---|---|
| 1 | Swayambhu | Yajna | Ajitha | Atri Marichi Angiras Vasishta Kra-tu Pula-ha Pulastya |
| 2 | Swarochisha | Ajitha & Tushita | Tushita | Rishaba, urja, Stamba, Niraya, kashyapa parivan, Datta |
| 3 | Outtama | Satyasena | Sudhama Shiva Japa & pratardana | Sons of Vasishta |

| | | | | |
|---|---|---|---|---|
| 4 | **Tamasa** | **Hari** | Satya<br>Hari<br>Sudhis<br>Swarupa | Agni<br>Prithu<br>Jyotir-dharma<br>Chaitra<br>Vanaka<br>Kavya pivara |
| 5 | **Raivata** | **Manasa** | Amithabha<br>Sume-dhasa<br>Vaikuntha<br>Bhuta-rajas | Devashri<br>Urdha-bahu<br>Hiranya-roma<br>Veda-bahu<br>Mahamuni<br>Sudhama<br>parjanya |
| 6 | **Chaksusha** | **Vaikuntha** | Lekha<br>Bhavya<br>Prithuga<br>Prasuta<br>Adya | Viraja<br>Sumedha<br>Madhu<br>Outtama<br>Havish-mat<br>Atinama<br>Sah(v)ishnu |
| 7 | **Vaivastha (The current manuvantara happening is completed with 50 years of current Brahma life time) i.e 155 trillion years** | **Vamana** | Vasu<br>Rudra<br>Aditya<br>(These 3 tribes consist of 33 demigods) | Atri<br>Gautama<br>Kashyapa<br>Vasishta<br>Vishwamitra<br>Jamad-agni<br>Bharadwaja |

| 8 | Savar(a)ni (Forth Coming) | Sarva-bhuma yet to happen | Mukhya Suthapa Amithabha | Galava Rama, Kripa Dipti-mat Vyasa, Drauni Rishya-shringa |
| 9 | Daksha-Savar(a)ni (Forth Coming) | Rishaba yet to happen | Para | Sabala Bhavya Vasu D(y)uthiman Medhatithi Jyotishman Satya |
| 10 | Brahma-Savar(a)ni (Forth Coming) | Vishwaksena yet to happen | Sudhama Vish(n)udhas Shatasankya | Satyakethu Aprati-mauja Havishman Sukrithi Apamurthi Satya Nabhaga |
| 11 | Dharma-Savar(a)ni (Forth Coming) | Dharmasethu yet to happen | Vihangama Kamagama Nirman-arathi | Agniteja Vapushman Nischara Aruni Vishnu Bhavishman Anaga |

| 12 | **Rudra-Savar(a)ni** (Forth Coming) | **Sudhama** yet to happen | Harita Sumanasa Lohita Sukrama | Sutapa Tapasvi Tapa-murthi Tapa-rati Tapa-d(y)uti Tapa-dhriti Tapa-dharna |
|----|----|----|----|----|
| 13 | **Rauchya** (Forth Coming) | **Yogeshwara** yet to happen | Sudharma Sukrama Sudhama | Nirmoha Nishprakampa Tatvadarshin Nirutsuka Dhritimat Avyaya Sutapa |
| 14 | **Indra-Savar(a)ni** (Forth Coming) | **Brihad-bhanu** yet to happen | Chaksusha Pavitra Bhrajira Va(v)riddha Kanishthas | Shuchi Agnibahu Magadha G(i)ridhara Shukra Yukta Ajitha |

The End of Cycle or Cosmic time similarly explained in Hinduism which is also closely correlating with the scienticfic model of Modern science (Modern Science Term used because validation given after Telescope Viewing or Pictures photographed from satellite).

| Hindu Cosmology<br>Vishnu Purana (Wilson, 1840) | Scientific Model<br>Schroeder and Smith (2008) |
| --- | --- |
| Extermination of human beings due to a prolonged hundred-year drought. | Life no longer sustainable with the increased solar flux. |
| Increased solar flux causes the Earth to dry up and the oceans and streams to boil over – a runaway greenhouse effect | A 10% increase in Sun's luminosity in about a billion years causes a substantial increase in the water vapor content of the atmosphere and the oceans start to evaporate. |
| A green-house effect causing an increase in Earth's surface temperature, setting the Earth on fire that turns it into a barren rock resembling the back of a tortoise. | An initial moist greenhouse effect will cause runaway evaporation until the oceans have boiled over; that much of the atmospheric water vapor will be lost to the stratosphere through ionization of water. |
| Incineration of the Earth and the space around it | |
| Formation of a giant stellar nebula that fills the sky with a myriad of colors - reds, blues, sapphire, jasmine, smoky - and shapes and sizes resembling giant elephant trunks, mountains, columns, and houses or hovels. | The subsequent dry greenhouse effect raises the Earth's surface temperature faster turning the earth into a molten remnant. |
| A hundred-year deluge that forms a giant cosmic ocean. | Earth, Mercury and Venus are engulfed by the Sun. |
| Dispersal of the stellar nebula by strong stellar winds that last for a hundred years. | A mild stellar wind<br><br>A small stellar nebula. |

<<<<<<< §µΨΦ∞ЖλπЦδ >>>>>>>

We dealt with the subject matter of creation and evolution as per Vedas and Modern Science has been compared and understood, on proceeding, we try to understand little deeper in understanding the branches of Metaphysics and Particle physics as we know Vedas do really deal with Particle Physics and Quantum Physics. Quantum Physics deals with the mechanism of change which is continuously happening and dealing with variations and reordering of subject matters connecting the Universe.

There are five main ideas represented in Quantum Theory:

1) Energy is not continuous, in small but discrete units.
2) The elementary particles behave both like particles and waves.
3) The movement of these particles is inherently random.
4) It is physically impossible to know both the position and the momentum of a particle at the same time.
5) The atomic world is nothing like the world we live in.

Pulling two quarks apart requires so much energy, that will create new particles in the process which attach themselves to the old quarks and thus it will never end up with free quarks to detect. The important point here is that the experiments that confirmed the existence of these quarks never actually detected them directly and no one really demanded such a proof. They saw certain particles coming in and certain particles going out according to what the theory predicted, but never actually saw the free quarks by themselves.

On the other hand, Vedas tells us about the existence of another type of particle called the *atma*, the soul, the self, having different properties from those of ordinary particles. This is not unusual in physics, as is the case with anti-matter particles. But unlike the ordinary matter and anti-matter particles, the *atma* has 3 main qualities: eternality, knowledge and bliss the cycle or evolution of atma. Scientifically the evolution of sub atomic particles. As per Vedas only eternality (soul or spirit which has the ability of moving out of body and getting back inside a body—Capable of moving to eternal worlds) and in Modern Science the same has been explained as Proton has the ability to evolve within itself as it is the only sub atomic particle which has the density of mass.

<<<<<<< §µΨΦ∞Жλπꚉδ >>>>>>>

Eternality can be analyzed as follows: the *atma* does not come into or out of existence, it is always present. It does not decay and degrade, it maintains its individuality always, it's infinitely small and it's not composed of more elementary particles. It's unchangeable, which along with the individuality property implies that it does not unite with other external and outside particles to create more complex structures, nor does it merge with others. The *atma* does not interact directly with matter, but given that the *atma* is the source of consciousness and the life force in a living entity, any entity displaying them reveals the presence of an *atma* in it and we deal this concept a bit in detail to understand better in next chapter.

This life of yours which you are living is not merely a piece of this entire existence, but in a certain sense the whole; only this whole is not so constituted that it can be surveyed in one single glance. This, as we know, is what the Brahmins (priests in the Vedas) express in that sacred, mystic formula which is yet really so simple and so clear; **tat tvam asi**, this is you. Or, again, in such words as "I am in the east and the west, I am above and below, I am this entire world."

This is a reference from Mundakya Upanishad mantra (above) in which the Vedic understanding of the connectivity of living entities is put forward to help the Bhakta (practitioner of yoga) to understand the difference between the body and the living entity. How the real nature of the living entity is realized only in union with the source, the supreme being (para-brahman/Narayana) through a platform of transcendental divine loving service.

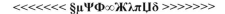

# 7

# *Sparks of the Idea from Vedas to Modern science*

The Matter we are going to deal now is comprehending the idealogy and mythological decoding, unraveling the secrets in universal activities and comprehending Metaphysics. **(RV) in brackets mentioned here in this chapter is denoting to Rig Veda.**

Anybody now will comprehend that the concept of Trinity in Hindu mythology were speaking Vishnu, Brahma and Rudra?!. The Basic Question why this 3 supreme deities holding the roles of Organiser, Creator and Destructor, In Vedas the order of Creation is mentioned as well. Is it what telling us what PROTON, NUETRON and ELECTRON is? Is this what Proton (Narayana), Neutron (Brahma) and Electron (Rudra)

We need to see what clarification does the particle physics is to offer us? Particle Physics?

The branch of physics that deals with the properties, relationships, and interactions of subatomic particles. Although the word "particle" can be used in reference to many objects the term "particle physics" usually refers to the study of the **fundamental** objects of the universe—fields that must be defined in order to explain the observed particles, and that cannot be defined by a combination of other fundamental fields. The Higgs boson has often called it "the God particle" because it's said to be what caused the "Big Bang" that created our universe many years ago. What is the Primary source? However I feel it is my duty to remind the readers that the recent discovery in CERN, Geneva, about God's Particle, the "PROTON QUARKS" is very old theory in vedic metaphysics and cosmological science.

I here take the opportunity to inform about Higgs boson, Peter Higgs, is a British theoretical physicist and emeritus professor at the University of Edinburgh and Boson.

Satyendranath Bose (1894-1974) Post graduated in mixed mathematics from University of Calcutta. The Research paper he presented later proved in creating the field of quantum statistics. Bose sent the paper to Albert Einstein in Germany, and the scientist recognized its importance, translated it into German and submitted it on Bose's behalf to the prestigious scientific journal *Zeitschrift für Physik*.

He worked alongside Einstein and Marie Curie, and many others. Einstein had adopted Bose's idea and extended it to atoms, which led to the prediction of the existence of phenomena that became known as the Bose-Einstein Condensate, a dense collection of bosons—particles with integer spin that are named for Bose.

According to July 2012 **New York Times** article release in which Bose is **described as the "Father of the God Particle'.** Several Nobel Prizes were awarded for research related to the concepts of the boson and the Bose-Einstein Condensate. **Bose was never awarded a Nobel Prize, despite his work on particle statistics**.

The Indian government honored Bose in 1954 with the title Padma Vibhushan, the second-highest civilian award in India. Five years later, he was appointed as the National Professor, the highest honor in the country for a scholar. Bose designated in that position for 15 years. Bose also became an adviser to the Council of Scientific and Industrial Research, as well as president of the Indian Physical Society and the National Institute of Science. He was elected general president of the Indian Science Congress and president of the Indian Statistical Institute. In 1958, he became a Fellow of the Royal Society.

According to CERN's statement the detection of the boson is a very "rare event— it takes around 1 trillion proton-proton collisions for each observed event, The "PROTON QUARKS". The primary source from Vedic metaphysical concept of cosmological science, THANKS TO INDIAN VEDAS?!!!!!!!!.

<<<<<<< §µΨΦ∞Жλπцδ >>>>>>>

The following mantra considered is higher most mantra in Vedas

**Gayatri mantra: Rig Veda: 3.62.10: Devanagiri (sanskrit)**

ॐ भूर्भुवः स्वः ।
तत्सवितुर्वरेण्यं ।
भर्गो देवस्य धीमहि। ।
धियो यो नः प्रचोदयात्॥ ।

**(actual Sanskrit hymns spelled in English Language with translated meaning)**

*Aum bhur (world of matter) **bhuvah** (world of life energies which sustains all life) s(u)vah*
*(world of mind) **Tat savitur** (Aditya-knowledge brighter like sun) **varen(i)yam** (excellent glory)*
***Bhargo** (glorious light of god's love and power) **devasya dhimahi** (may we attain)*
***Dhiyo**(intellect) **yo** (who) **nah** (ours) **pracodayath(u)** (request from supreme god).*

<<<<<<< §μΨΦ∞ℲλπЦδ >>>>>>>

**Yajur Veda Chapter 1: Sukta 3: (actual Sanskrit hymns spelled in English Language)**

**Vaso: pavithramasi shathadhaaram vaso: pavithramasi sahasradhaaram**
**Devastwa savithaa punathu vaso: pavithrena shathadhaarena**
**Supwaa kaamadhuksha:**

**Persipicacious Meaning**: Savitr, the one who creates the prosperity and power in hundredfold, let you amplify it to thousand fold stream of the same. The lord who showers the collective energy on you, make all your goals realized and make them purified. Let all your journey ways be sanctified.

Yajur Veda Chapter 1: Sukta 12: (actual Sanskrit hymns spelled in English Language)

Pavithre (s)sto vaishnavy(o) savithur vah: prasavauth punyaamya-
Chhidraena pavithrena sooryasya rashmibhi: deveeraapoo agraeeguvo
Agraeepuvooyo gra imamadya yagnam nayathaagrae yagnapathing
sudhaathum yagnapathim devayugam.

**Persipicacious Meaning**: Oh the celestrial waters that reside in the heaven and the earth, you belong to the omnipresent god. With his intention, I use the rays of the sun to purify you. Not only it flows forward, but also brings in the purity with the flow. Not only the conductor of the yagna is enabled to move forward, but also make the yagna itself to move forward similarly. This yajamana who is having good traits is also loved by the Devas.

Yajur Veda Chapter 2: Sukta 2: (actual Sanskrit hymns spelled in English Language)

Aadithyae vyundanamasi vishnau stuposyoornamradasam twaa
Strunaami swasastaam devaebhyo bhuvapathayae swaha
Bhuvanapathayae swaha bhoothaanam patayae swaha

**Persipicacious Meaning**: You (Vishnu) is the one who provides water to the earth, with yagna you are being crowned, let me cover you with darbha grass, the dharba grass which I am using to cover you again and again; its all soft, let you reside on all folks who have the divine wisdom, my salutation to the celestial owner of the lands and the houses, my salutations to celestial owner of the five subtle elements.

Physical sciences relating to agriculture, medicine, astronomy mathematics particularly algebra, etc. are described **in Rig Veda.1-71-9, Rig Veda 4-57-5, Sama Veda 121. This is what Vedas scriptures mentioned is exactly matching with Particle Physics (Physical Sciences)?!**

**SAMA Veda 222** describes the same scientific truth as "He Keeps his wonderful form in every atom, spreads and occupies earth, the sky and the layer of space in between".

<<<<<<< §µΨΦ∞ЖλπЏδ >>>>>>>

**Atharva Veda 2-5-12** refers to two kinds of electricity—positive and negative along with its friendly and destructive use. The electricity is hidden in water and when it comes out, it spreads light and provides energy**(Rig Veda 1-16-5)**. Its use in weapons were mentioned **(Rig Veda. 1-85-5)**. Because of heat energy in the electricity, there is need to have various precautions against electricity. For electricity the words used in Vedic language are "*Viduat Raksha*." *Viduat* in Sanskrit is electricity and *Raksha* is protection. Agni also contains energy and electricity **(Rig Veda 1-45-5)**. Electricity protects people and should be used as destructive energy against wicked persons and enemies with the help of weapons, which work on electricity **(Rig Veda 1-86-9)**.

*Viman diye neshu* for vehicles like aircrafts **(Rig Veda 1-34-1, 9 and Rig Veda 1-85-7)**. A few hymns in **Atharva Veda (1-11-1 to 7)** are devoted to childbirth like, foetus is surrounded by natural elements that move the child in the womb and prepare the woman for giving birth.

Atharva Veda which deals with full of Human Physiology in detail in inward and outward repercussion, Psychology, Diseases, Household maintanance, Field of Agriculture, Cattle, Business and Gambling

**Atharva Veda Chapter 4: Suktha III, 23. Process for conceiving a son (pumsavanam).**

1. That which has caused thee to miscarry do we drive away from you, that very thing do we deposit outside of thee, away in a far place. Into thy womb shall enter a male germ, May a man be born there, a son ten months old. A male son do thou produce, and after him a male shall be born! Thou shalt be the mother of sons, of those who are born.

**Atharva Veda Chapter 4: Suktha VI, 11. Explaining the process for conceiving a son (pumsavanam).**

1. The **asvattha (ficus religiosa)** has mounted the **samî (mimosa suma):** the elixir of the species in right proportion in mixture shall produce the sperm fertility shall result in producing a male child, is the way to obtain a son; that a man fertilises to a woman for pregnancy. a**svattha or (ficus religiosa)** Ficus religiosa is actually A Fig is a species of fig native to India, Bangladesh, Nepal, Pakistan, Sri Lanka, south-west China and Indochina. It belongs to the Moraceae, to mulberry tree family. It is also known as the Bo-Tree or Peepal or Pippal m**imosa suma** or (Acacia polyacantha) also known as **White Thorn** is a flowering tree which can grow more than 20 metres

tall. The tree is native to Africa, India, the Indian Ocean and Asia, but it has also been introduced to the Caribbean.

**Atharva Veda Chapter 1: Suktha VI, 109. The pepper corn a cure for wounds.**

1. The pepper corn cures the wounds that was caused by fire injury, it also cures the wounds from knife cuts and knife stabs. Powerful to secure the life back, this plant shall be . . . .

2. The pepper corns bind to one another, as they peeled out, yet it still retains its virality: on finding this alive, that man shall not suffer harm . . . .

3. The Asuras did dig you into the ground, the gods cast you up again, as a cure for disease produced by wind (in the body), moreover is a cure for wounds struck by fire wounds . . . .

**Atharva Veda Chapter 1: 17. Process to stop the flow of blood.**

1. The maidens that go yonder, the veins, clothed in red garments, like sisters without a brother, benefit of strength, they shall stand still . . . .

2. Stand still, you lower one, stand still, you higher one; do you in the middle also stand still! The most tiny (vein) stands still: may then the great artery also stand still! Of the hundred arteries, and the thousand veins, those in the middle here have indeed stood still. At the same time the ends have ceased

3. Around you has passed a great sandy dike: stand still, pray take your case!

**Shukla yajur veda 3.54, Rig veda 10.57.4:**

(preaching by rishis to students) The spirit shall come to you to body again for wisdom, energy and life that the subject matter may long behold the sun.

What is that Ancient Intellectuals have been guiding us to is that lead for modern discoveries in the area of medicine, bio genetics, DNA Reengineering etc., Ancient Intellectuals considered the practice of medicine as an art mostly mixed up with science,

These Vedic Hyms in the scriptures speaks itself for its expositions of scientific facts involved into the religion as indians consider it as ancient science, Ancient Facts of the vedas were not falsified till date or not changed from its original explanations, it told what exists, how and what evolves, the reasons and causes, the results, the achievements done etc . . . ., is still intact and the more research done, the more clarity on the hymns still evolving to the the understandings.

<<<<<<< §μΨΦ∞Жλπυδ >>>>>>>

We will certainly see the other subsect-religions birth, success, sustinance and fall and its part in India has clearly indicated the domination by hinduism. Other Religions of India have branched out on its own Philosophies from Hinduism framed their own way of discipline and cult followeres of the Sect idealogies. Few to be mentioned were BUDDHISM, JAINISM and SIKHISM.

<<<<<<< §μΨΦ∞Жλπυδ >>>>>>>

**Here in the following paragraphs i use the terms Saints based religion and scholar based religions.**

**Saints liberated themselves from the family or remained unmarried and devoted themselves to the divinity and philosophy and went on writing vendantas(philosophical commentaries on Vedas) and established vedic schools which taught Vedas and ancient wisdom. but Scholars lived in a family system had wife and children and continued to develope the followers base by ideating the teachings of the cult worship for the society about the theory of social stature and communal wisdom and continued their services to the promotion of cult services being the motive,**

<<<<<<< §μΨΦ∞Жλπυδ >>>>>>>

JAINISM: **A Saint based Religion**, Jainism dates to the 6th century BC in India. The religion derives its name from the *jina*s (conquerors), a title given to twenty-four great teachers, Mahavira, the last of them and is considered the founder of Jainism. The ultimate goal of Jainism the liberation from rebirth, which is attained through the elimination of accumulated karma which is also the highest principle in Hinduism is "MOKSHA". Like Hinduism, Jainism also believes in multilayered universe containing both heavens and hells. Jainism doctrines emphasizing a peaceful and disciplined life. These principles include non-violence in all parts of life (verbal, physical, and mental), speaking truth, sexual monogamy, and the detachment from material things. Jains typically are strict vegetarians which restricts the sorts of occupations the may follow (no agriculture and farming, because insects are inadvertently harmed in plowing).

BUDDHISM: **A Saint based religion**, Buddhism had seen a steady growth from its beginnings in the 6th century B.C to its acknowledgement as religion of the Maurya Empire under Ashoka in the 3rd century B.C. It continued to flourished and had its prominence for almost 1000-1200 years, and spread even beyond the Indian subcontinent to Central Asia and beyond to China. **Decline of Buddhism in India**, the land of its birth, occurred for a variety of reasons and happened even as it continued to flourish beyond the frontiers of India. Decline set to happen during the later Gupta era and under the Pala Empire. Chinese monks travelling through the region between the 5th and 8th centuries, such as Faxian, Xuanzang, Yijing, Huisheng, and Song Yun, began to speak of a decline of the Buddhist *sangha*, Further Decline continued after the fall of the Pala dynasty in the 12th century CE and gradual Muslim conquest happened in the Indian subcontinent.

SIKHISM: **A Scholar based religion** is founded during the mid 15th century in the Punjab region, by Guru Nanak during Muslim rulers conquest and continued to progress through the ten successive Sikh gurus (the last guru being as per the scripture is *Guru Granth Sahib*). This system of religious philosophy and expression has been traditionally known as the **Gurmat** meaning **Knowledge of the GURU**. Guru Nanak, Sikh Guru established the system of the Langar, or free kitchen, designed to safehold equality between all people and express the ethics of sharing, community, inclusiveness and oneness of all humankind. In addition to sharing with others, Guru Nanak encouraged earning/making a living honestly without exploitation or fraud and also doing meditation on God's name or qualities. Guru Hargobind, the 6th Sikh Guru, established the political/temporal (Miri) and spiritual (Piri) realms to be mutually coexistent According to the 9th Sikh Guru, Tegh Bahadhur, the ideal Sikh should have both Shakti (power that resides in the temporal), and Bhakti(spiritual meditative qualities). This was developed into the concept of the baptised saint soldier of the Khalsa by the 10th Sikh Guru, Gobind Singh. In Sikhism, God termed *VAHEGURU* is shapeless, timeless, and sightless (unable to be seen with the physical eye), signifying the universality of God. It states that God is omnipresent and infinite with power over everything.

Jainism, Buddhism and Sikhism were religions considered as the religions of Saints and Scholars who had ideal visions, The followers of both the religion accepted and acknowledged Saints and Scholars as demigods. The cult developed of worshipping scholars as demigods even got spilled over to other religions as well.

<<<<<<< §µΨΦ∞Жλπƕδ >>>>>>>

Even today there is a separate group of community found common from all the religions ended up following Sai Baba, this particular group were still not clear about sai baba's birthplace or about their parents or how he learned the philosophical ideas and the medicinal treatments, or from which GURU is not clear, as he was known to be a wandering scholar singing the songs in praise of Lord Rama (Incarnation of Lord Vishnu) and stayed mostly in the mosques of Islam Religion. The life period of this scholar to be said after the fall of Muslim dynasty in india, when india is under british rule. Even today in the most parts of india there are temples to this scholar is built and while closely noticing the poojari (the chanter of prayer) at his shrines doesn't chat any prayers during poojas, or found chanting prayers of a different deities, and it is found the usual chanting on this religious scholar as the prayer songs in the name and praise of the scholar himself in chorus as bhajans.

As these followers of GURUs and worshippers group majorly identified with non vedic believers and great propagandas done among other communities and seemingly to operate in franchise groups found collecting the donations and building up shrines to develop the cult worship and completely failed to acknowledge the meanings of Vedas and knowledge in siddha and ayurvedic treatements he explained and performed and cured many of them during his days.

<<<<<<< §µΨΦ∞Жλπƕδ >>>>>>>

**Awakening and establishing the foothold of Hinduism and gaining momentum in India again**:

During 8th Century the famous South Indian religious saint and scholar started on his voyage to challenge the Buddhist and Jain scholars on religious mythological meanings and dimensioned the strength of Hinduism to explain on the ideologies to its survival and defeated the other religion scholars in debates, countering them using their own philosophical ideas and explained the deviations and variations in the philosophical ideas which is not logically, technically or fundamentally possible, as it continued through out india, scholars after scholars of many religions at that time in india flourishing were defetead in debates by Adi Shankaracharya and the religious literary and discourse work done by him is **Advaita vedanta**. After his life time works in philosophies followed him few more famous hindu religious saints from south india, renowned for their vedic knowledge and discourses in vedantas, puranas and Upanishads by Ramanujacharya, Madhvacharya and Nimbarkacharya shedding more clarity on hinduistic religious principles of the predecessors and establishing religious schools all over india teaching religious philosophies and vedas all over india, which ultimately gained momentum back to hinduism back again in India.

The Advaita Vedanta focuses on the following basic concepts:

Para-brahman or Brahman(Supreme and Spiritual Realm), atman(soul), vidya (knowledge), avidya (ignorance), maya(illusion), karma (accountability of fate & destiny) and moksha(liberation).

Para-brahman (Supreme and Spiritual Realm) is the Ultimate, Erudite and Spiritual Acceptance. Brahman is eternal. Brahman is beyond gratification. It is beyond noumenal and forms. Brahman can not be perceived nor could it be described by words. It is beyond senses and intellect. It is indefinable.

Brahman (the material realm) can be considered as pure consciousness In Vedanta philosophy, the svaroop (form) of Brahman is referred to as Sachchidananda. Brahman is Sat-Chitta-Ananda (Pure Existence-Pure Consciousness-Pure Bliss). Brahman is eternal, redundant and unthinkable pure-existence, but it is not the Para-brahman.

Atman (soul) is the inmost Self or Spirit of man but different. Atman is the fundamental, ultimate, eternal, redundant pure consciousness. It is well explained that Atman belongs to Brahman and inturn to Para-brahman. Brahman is the ultimate reality behind all world-objects and Atman is pure spirit in all beings. Both Brahman and Atman are not different realities. They are identical. For practical purposes, they are referred to separately, which they are not. They are the eternal, perpetual realities underlying all existence. They are two different 'fervent' for one and the same reality behind all the objects, all matter, all beings of the universe. In simple words the attributes of mind and soul.

Maya (illusion) is the unique power (shakti) of Brahman. Maya three gunas or attributes. But Shuddha Brahman is nirguna and is free from attributes. **Shuddha Nirguna Brahman alone is the Supreme Reality**. When **Nirguna Brahman comes to acceptance without protest and acknowledges the attributes of maya**, Saguna Brahman (Tridev) is God, the creator, sustainer and destroyer of the world. **Saguna**

**Brahman is Ishvara or a 'personal god.' Man worships gods in different forms and names**.

Brahman manifests itself in the world with the help of Maya. The worldly objects come into life due to the power of maya. Maya and its creation is termed illusionary. It does not mean that the world is not real. Unreality and illusion are different. An illusion may not be an unreality for an illusion is grounded in reality. Reality is that which exists on its own. Maya is dependent on Brahman. Maya has created the world of appearances. So the world is illusion. But this does not mean at all that the world is non-existent. The Advaita Vedanta, with the help of the famous "rope-snake" illustration from scripture, maintains that it is neither ultimately real, nor wholly unreal, illusionary and non existent.'

Avidya (ignorance) has its seat in the human intellect. Avidya means not only absence of knowledge, but also erroneous knowledge. A man trapped in ignorance does not know what is real and thinks that the appearances are real. An individual identifies himself with empirical self. He equates his existence with the physical body. Under the influence of Maya and Avidya, he dis-associates himself from the ultimate reality. When the man acquires knowledge, the duality of the self and Brahman disappears. He realizes that the self is really one with Brahman. This realization of the self puts an end to the ignorance (avidya).

Moksha (liberation) is freedom from bondage of ignorance. Man suffers in the grip of incessant desires and ignorance. Upon realization of the self, one becomes free from the shackles of desires, aspirations, passions, karma and avidya. This is Moksha (kaivalya) or liberation. Moksha is to be attained here and now during this life-span only.

Knowledge and Truth are of two kinds: the lower one and the higher one. The lower, conventional knowledge and truth is referred to as vyavahrika satya. It is a product of the senses and the intellect. The higher one is referred to the paramarthika satya. It is absolute. It is beyond words, thoughts, perception or conception. It is in no way, related to the senses and the intellect. It is non-perceptual and non-conceptual. It is a product of sublimal rendition and "divine vision". The higher knowledge and truth brings about radical transformation in an individual so it is soteriological (Salvation theory).

**Visisht(a)advaita (By Sri Ramanujacharya)** is also an Advaita with Sri sampradaya, since only God the Absolute, omnipresent Self exists. However, **Sri**

**Ramanujacharya** concept of God refers to that Supreme entity which contains all within itself, the entire universe, including all living beings, are fundamentally real and internally distinguishable from one another.

However, there is only one total reality, as God includes all existence within Its very being. The individual self and the universe exist as god's attributes, since God pervades absolutely everything and gifts these substances with their reality. In other words, God is the indwelling. Self of all, and this ALL is real as they are included in his body. Therefore, Visishtadvaita literally means non-duality of the qualified, since God is qualified by innumerable glorious attributes, including individual selves and matter.

Adi Sankaracharya commented Liberation is eternal communion and service of God, the supreme, infinite, blissful Absolute.

Sri Ramanujacharya commented that such liberation is achieved by constant meditation on God's supreme perfections. His omni—presence, His splendorous forms, His actions in his various manifestations, His existence in the hearts of all creatures, His nature as the first cause of all. such meditation, when practiced with a pure heart and mind and filled with extreme love, will yield a better and better conception of God in the mind of devotee over time, eventually leading to recollection of God, so vivid it is like sight itself. Such a vision, when practiced to the point of being unbroken, is the liberating knowledge spoken of in the vedas, a result of God's love of his beloved devotee.

## THE DIFFERENCE IN APPROACH: ADVAITA AND VISISHT(A)ADVAITA: EXPLAINED:

The **Adi Shankaracharya's (Advaita) is absolute monotheism** conception of the Absolute has no attributes. Hence the discipline of meditation there does not rely on bhakti (devotion) in the end. **Then Sri Ramanujacharya's Visisht(a)advaita has as its centerpiece a supreme being full of perfections and attributes, so the individual aspirant has no choice but to reveal in these kalyana gunas (limitless and boundless).** From this basic difference the triad of Para-brahman (Supreme soul) along with the universe and the sentient beings is Brahman, which signifies the completeness of existence. **Para-brahman (Supreme realm) endowed with innumerable auspicious qualities (Kalyana Gunas). Para-brahman (Supreme realm) is perfect, omniscient, omnipresent, incorporeal, independent, the** creator of the universe, its active ruler and also its eventual destroyer. Visisht(a) advaitha is considered as Qualified Non-Dualism.

As the flourishing of Vedas happened from the ancient time to present day there is still the evidences of events occurred and lead to the **Dark Years and the era of discontinuance**, as anything and everything being discontinued while the people cared about only being alive and struggled for survival and this part of the dark ages actually help us to obtain the interesting view point of unlocking the clues of major transitions in earth and survival of Vedas across these dark years which actually continued for about millenniums. We dig into this theory deep in the proceeding chapter . . . .

<<<<<<< §μΨΦ∞Ж λπЦδ >>>>>>>

# 8

# *The Dark Years and Discontinued era*

In this chapter here we will try and comprehend the missing information and the dark Years Plunging down in humanitarian beliefs and raging violence across civilisations across the world continuing to the very day.

**EGYPT**: The initial breakdown of the old civilisation was caused by a sudden, unanticipated, catastrophic reduction in the Nile floods over two or three decades. This was so severe that famine gripped the country and paralysed the political governance. People were forced to commit of unheard atrocities and evil acts such as eating their own children and violating the sacred sanctity of the royal dead.

The failure of the floods is shown by the fact that the Faiyum, a lake approximately 45-60 metres deep, dried up. This means that the lake actually evaporated over time. These low floods were related to global climatic warming which reduced the amount of rainfall in Ethiopia, North and East Africa. This correlates with a shift to drier climate in south-eastern Europe c.2200-2100 BC. Also, the reappearance of oak at White Moss, UK, suggests fluctuating wetness in around 2190-1891 BC. In Italy, drier conditions are found around 2200-1900 BC in Lake Castglione. Dry spells have also been detected as far away as Western Tibet at Lake Sumxi.

Long-term variations in Nile floods are beyond the perceptions of people. The Nile during the prosperous times of the old civlilisation, is regarded unquestionably as the source of life in Egypt. Therefore, the Nile can be considered as the force which destroyed the civilization that it had nurtured. Inconceivably as it might be, the Nile is a temperamental river. The volume of flood discharge varies wildly in episodes which range from decades to hundreds of years. Furthermore, there is the impact of disastrous years where the floods can be unbelievably low or high.

People emigrated in crowds and that those who remained habitually ate human flesh; parents even ate their own children. Graves were ransacked for food, assassinations and robbery reigned, unchecked and noblewomen implored to be bought as slaves.

There are four successive episodes during this upheaval of Egyptian civilisation. First came the initial episode of shock, upheaval and fragmentation which were caused by low floods. This lasted from the end of the 6th Dynasty to the end of the 8th (perhaps as early as 2100 and certainly by 2155-2134 BC). Then came the second episode of rehabilitation and re-development of regional politics which commenced 2134 BC. This encompassed the first two generations after the end of the 8th Dynasty (the 9th in Herakleopolis) and the first part of the 10th in Thebes. This was followed by the struggle between Thebes and Herakleopolis during the reign of Antef, who succeeded in re-establishing order during his 50-year reign. This incidentally did not lead to any weak successors. Finally occurred the consolidation of national unity by Mentuhotepe II and his immediate successors after c.2020 BC.

**INDIA**: In a supplementary chapter of the Vajasaneyi-Samhita of the Yajurveda (34.11), Saraswati river is mentioned in a context apparently meaning the Sindhu: "Five rivers flowing on their way speed onward to Sarasvati, but then become Saraswati a fivefold river in the land." According to the medieval commentator Uvata, the five tributaries of the Sarasvati were the Punjab rivers Drishadvati, Satudri (Sutlej), Chandrabhaga (Chenab), Vipasa (Beas) and the Iravati (Ravi). Saraswati river now has no traces of once a river contributed a fertile land and civilization in india.

During the Pleistocene period the Himalayan mountains were under glacial cover and climate was fluctuating between glacial and interglacial phases. Around 35,000 yrs BC approximately, the present Thar Desert did enjoyed the wet climate and green valleys. River Saraswati (also known as Saraswati Nadi, Saraswati Nala, Sarsuti and Chautang in certain places, variously spelt as Sarasvati) is believed to have flowed during 6000-3000 BC from the melting glaciers of Garhwal Himalaya to Arabian Sea through the Thar Desert. Several researchers agree about the existence of palaeochannels. According to the Ground Water Cell of Haryana, a large number of water wells fall on these palaeochannels and their lithology is coarse sand or gravel of riverine nature. Now palaeochannels exhibit discontinuous drainage.

It does seem to have been a major contributory factor, but probably not the only one. It is certain that the urban Harappan sites in the Sarasvati basin, such as Kalibangan, Banawali or Rakhigarhi, had to be abandoned. In the Indus basin, on the contrary, floods and consequent shifts in the Indus appear to have occurred. All this must have greatly impacted agricultural resources and possibly the urban administration. The Harappan state was geographically quite overstretched, from the Yamuna almost to Iran, and from northern Afghanistan to the Narmada; it apparently could not survive these upheavals, and the Harappans had to revert to a rural stage.

The explanation offered by most scholars, geologists in particular, is that the Sarasvati was partly fed by waters from the Sutlej (in the west) and the Yamuna (in the east). Indeed many palaeo-channels connecting those three rivers systems have been traced. Now, the watershed between the Yamuna and the Sutlej is a very flat and seismically active region; it has been proposed that it underwent a slight upliftment, which drove away the Sutlej and the Yamuna, leaving the Sarasvati with only a few seasonal tributaries originating in the Shivaliks.

**CAMBODIA**: **Angkor Wat** is a Hindu temple complex in <u>Cambodia</u> and the largest religious monument in the world. The temple was built by the Khmer (Ruler) King <u>Suryavarman II</u> in the early 12th century in <u>Yasodharapura</u>. Breaking from the Shaivism tradition of previous kings, Angkor Wat was instead dedicated to Lord Vishnu. As the best-preserved temple at the site, it is the only one to have remained a significant religious centre since its foundation. Prior to this time the temple was known as *Preah Pisnulok* (Vara Vishnuloka in Sanskrit). The flooding of Mekong river have destroyed once flourished civilization from 8th century AD to 14 Century AD at Cambodia, isolating the the famous Angkor Wat monument temple into the dense Rain-forest, A complete civilization lost due to flooding, in contrary to the civilization lost due to the river dry-ups.

THE DARK YEARS contributing to the death of the river valley civilization and the migration of ancient communities toward the high lands or mountain region and to the other water resources, changing the belief systems, culture, eating habits and changing the colors of the culture altogether. The dark years contributing much to the rise of new cults and worship systems overlapping the old.

Perhaps the only river dry ups and river floodings alone could create this kind of lengthy dark ages, I don't think so, historical evidences are pointing to geological disturbances like Earthquakes, Volcanic eruptions, forest fire also fueled the death tolls along side the dark years which lead to the era of discontinuance. The era of discontinauance forced civilizations to give mixed emotions among people either creating extreme fear on nature and the strong feel of disbelief in the concept of god due to the kind of loss and grief being suffered, creating parallel chaotic situation which eventually ended up in leading to the fall of dynasties, kingdoms, civil wars, lootings, murders etc.,

The recently popularized Mayan civilization cannot be ruled out as well, from the era of discontinuance as it seemed to be the last known civilization (not religion) to be uprooted for further era of discontinuance reasoning the european invaders for expansion of colonization. Mayan history can be characterized as cycles of rise and fall, rose in prominence and fell into decline, only to be replaced by others. It could also be described as one of continuity and change, guided by another civilization(olmec civilization) that remains the foundation of their culture.

<<<<<<< §µΨΦ∞ЖλπЏδ >>>>>>>

The discontinuance in the historical recordings all long analysed was not found in the sub continent unlike the fall of Egyptian civilization and Indus valley civilization disconnecting the afganistan from Indian map as a result of dry up of river saraswati but still remained the matter of discontinauance did not happened in Indian sub-continent, instead it was spreading continuously to the south east asian region and knowingly to the islands near Africa too, what is really causing this habitats moving to the different land part from Indian subcontinent, is this happened because of growing population, as this reason being logically sound, but it did not seemed to be the survival reasons alone, rather the religion of Hinduism growing stronger, it seemed to the inhabitants and settlers were teaching the existing tribal people and nomads about the culture and civilization, as it is still not looping to the conclusion point of view,

There is some thing indicating far further to the demolition of trade practices, the loss of value to the precious metals, people were forced to kill themselves to avoid further grief and pain of survival difficulties, as we are looking for the obvious angle of notion is not yet reveting the picture of the important time cycle of dark years and discontinuance.

<<<<<<< §µΨΦ∞ЖλπЏδ >>>>>>>

This particular notion and fact of melting up of Ice in North pole and south pole struck me intensely to rethink about the the causes and reasons, the scientists were crying about El-nino and El-Nina, but what is it really? Sea surface temperature in the equatorial Pacific Ocean (above). El Nino is characterized by unusually warm temperatures and El Nina by unusually cool temperatures in the equatorial Pacific. Anomalies represent deviations from normal temperature values, with unusually warm temperatures and unusually cold. Something struck to my mind like thunderbolt, what we dealt earlier in the 2nd chapter about history repeating itself in different dimensions?

The most fertile lands getting dried up with constant reduction in rainfall points to the heat wave on surface of the earth should have caused the melting of ice from north pole (the present ARCTIC REGION) and south pole (ANTARTICA) as the nature tried to balance the climate from its own resources, which is actually compensated the loss of water level on earth resulting in rising ocean level, which keeps me guessing the unprecedented change happened of massive devastation, but thinking upon there seems to be something missing in the blocks which is simply not falling into the place and not giving us the picture perfect. The only source left is going back to the Indian Vedas for reference and confirmation.

<<<<<<< §µΨΦ∞Жλπџδ >>>>>>>

Again decoding the Vedas and ancient vedic scriptures, for the tips of historical accounts once again giving us the lending hand of its recordings of major events on earth in its Puranas and Upanishads, What is it really happened, decoding the concept of Matsya avatar as we know it is the first of 9 avatars of Lord Narayana (Vishnu) so far, giving us the hints of world getting flooded long ago and life beings recycling to its life on earth, lets try to conceptualise in understanding the essence of matsya avatar from Vishnu Purana,

The VEDAS, with the aid of which, BRAHMA, the creator God, performs his role, happened to slip out of his possession at a moment of his respite. An ASURA (Enemy of God) who is alert, observed it and immediately stolen them. But, Vishnu (Lord Narayana), the preserver-God, was watching this. Since PRALAYA—Dissolution—was to follow soon, the VEDAS would be lost for the next spell of Creation, unless they were retrieved.

As God was wondering what was to be done, he noticed Sage Satyavrata, who was doing penance subsisting on water alone, making the ritual offering to water of God. God immediately assumed the form of a Fish. As satyavrata scooped water from the flowing river, he saw a tiny fish in the water he had scooped. When he tried to put it back into the river, the Fish entered the sage's bowl and requested the sage not to do so as it would be eaten up by the big fishes in the river. Satyavrata took pity and took it into his KAMANDALU (an oblong water pot made of a dry coconut) and went back to his hermitage. Overnight, the Fish grew too big to be in the Kamandalu.

The next morning when the sage looked it up, the Fish requested him to remove him to a larger vessel. Satyavrata did so but soon the fish became too big for the larger vessel also. Addressing the Sage, the fish said that he should protect him and find a suitable living space for it. Satyavrata then emptied the vessel into a large pond near the hermitage along with the Fish. But, in no time the Fish grew as large as the pond and filled it. Then the Fish exhorted the Sage to take it to a large and deep lake. Although Satyavrata took it to several lakes, one larger than the other, the Fish kept growing and bigger and bigger. It went on asking for larger and larger living space.

Satyavrata got vexed and decided to put it into the ocean. While he reached the ocean, the Fish addressed him thus: "oh Sage, do not put me into the ocean, I am sure to be swallowed by the gigantic creature there." Satyavrata became suspicious now. He was despairing to know how a fish could in a day grow so big as the largest lake and still find it not big enough for it to live there. In a flash, he realised that it was Lord Vishnu in the form of Fish. Satyavrata immediately prostrated before the Fish and prayed to be told why God had appeared in the form of Fish. The Lord told his devotee, "Just a week from now, the ocean will rise and inundate the entire Universe for the dissolution of creation. At that time you will see a spacious boat approaching you. Do collect all the seeds, plants and animals required for the next spell of creation and get into the boat and await me. Take VASUKI, the kind of Serpents, also with you. The SAPTHA RISHIS (seven Sages) will also be with you".

The Fish left for fulfilling its mission. Hayagriva saw the gigantic fish approaching him and was overtaken by fear. He held the VEDAS tightly in his mouth. But soon the Divine Fish slew him and recovered the VEDAS and restored them to BRAHMA for him to resume the function of CREATION at the appropriate time.

As foretold by the DIVINE FISH, PRALAYA (deluge) set in and on the turbulent waters, a boat appeared. Sage Satyavrata, the Saptha Rishis and all the living creatures found safety in the boat. The Fish in the colour of gold and now with a horn came by and ordered that the boat be tied to its horn using VASUKI as a rope. While the boat floated safely on the rising and enveloping waters, Lord VISHNU taught the Rishis the highest form of truth(Realm). This collection of truth(Realm) has come to us in the form of MATSYA PURANA.

(Meaning to be understood from the above paragraph as: As nature himself as god and god himself in nature being the fundamental principle in Hinduism, is testing Sage Satyvrata's ego, will and devotion simultaneously, as sage satyvrata gets vexed of tiny fish's pace of growth (nature act of balancing) to lash out of human's egoistic nature, Nature giving the tip to sage satyavrata to realize and surrenders to the force of nature, nature instructing to collect and preserve all the seeds of plants and animals (so that the DNA of male and female species of the same animal could be protected and shall reproduce the cycle of births) proving the idea, none of the properties of creation and the nature's balance should not be tested for severance, as humans could not survive without ecological system is the lesson given to sage satyavrata and to us to understand. Hayagriva is the horse headed saint who stole the vedas, Horse headed? Horse were known for its intelligence level and the sense to react immediately to the nature scientifically proven level of memory power is very high compared to other animals, designed in its DNA but still an animal. Lesson to human, dangers involved in possessing the Vedas with the wrong man's brain with high capacity memory power with highly scientific knowledge) is unpredictable danger.

This story is closely similar to the <u>Noah's Ark</u> story in the <u>Hebrew</u> <u>Book of Genesis,</u> perhaps reminding us of the fact that most early civilizations were established near large rivers which were prone to frequent flooding.

<<<<<<< §μΨΦ∞ЖλπℲδ >>>>>>>

What if this Elnino and Elnina happened long ago like millions of years ago, What if the massive meteor crash landed on earth which is now standing tall as mountain range since then, with all possibilities in mind?!! what if all this happened in chain reaction as one reason leading to the another cause and so on

A meteor crash with El nino and El nina with Volcanic eruptions and Polar Ice Melt with forest fire, leaving one part of the fertile land to change itself to large deserts and leaving a large part of land under ocean to submerge. Shifting and recycling of water theory of evoporation to rainfall.

The fundamental principle of nature is the design of masterpiece crafted by god as per Hinduism which runs in cyclical paradox of each and every single species including the humans, inside the larger cycles of paradox of every single atom (atman-soul) in this earth is getting recycled into the the nature of larger paradox i.e. governing paradox (paramatman—Supreme soul or Supreme realm).

As we are stepping closer in unlocking the secret of the dark years and discontinuance, the darker years always gives the grid lock making it difficult to understand the operational logics involved in it. We will try and understand in what proportions the designs realigns itself to regain its stability on balancing itself, the nature

Understanding the higher theory of metaphysics, one should have the doubt raising in his mind that whether these nature's phenomenon is only applicable to life beings on earth or is it applicable to the earth itself and to its underground properties?

Yes it is applicable to earth too as we saw the properties of the earth on the surface and the properties below the earth are one and the same, but in different forms, if earth's surface undergoes the changes and the earth and its underground properties too will have the same reasons and causes. Vedas again showing it, and Hinduism explains it in puranas.

<<<<<<< §µΨΦ∞Ж λπЦδ >>>>>>>

As we are heading to the fantastic journey of unlocking this secret will emerge as key to support all the notions, perceptions, ideas and the logics, in the next chapter, we will be unlocking the secret step by step with awesome surprises and notions, Enjoy the chapter 9—unlocking the secret what we are looking for . . . .

# 9

# *Unlocking the Secret*

While analyzing deeper and deeper on what may have caused the dark years and era of discontinuance on that missing point of view, something is clearly understandable, the massive tectonic events due to earthquakes cannot cause the massive land part to shift as it is, on consulting the geological experts from a reputed university in India, hinted a clue, how part of Dwaraka city, in india submerged under water after Gurukshetra War, after the massive explosion of Brahmahastra, the massive Hyderogen atom bomb explosion is mentioned in Bhagavad-gita, as, the bright light was so powerful it equaled the brightness of 1000 suns, the kind of heat it created left many people disoriented and buried in ashes later found in geological research about lost city of Dwaraka, a huge graveyard with thousands of dead bodies found scattered, looking at the structure of the submerged city it was understood later to be the streets of Dwaraka not the graveyard. Similarly, the lost city of poompuhar, near Mahabalipuram the temples got eroded by raising sea level, near Chennai city at Tamil Nadu.

Perhaps an obvious point of perception raised, as this heat due to the explosion should have dried up the saraswati river, from Himalayas to gujarat. This explosion is bigger in devastation scale occurred in land. On understanding there are few tectonic plates movement found but it is highly impossible this movement cannot possibly carried the continents all together to a different geographical locations.

<<<<<<< §µΨΦ∞ℑℵλπЏδ >>>>>>>

It should have been the submerged and the submerge of Lemuria (Continent) underwater due to water level raise. Oral tradition in Polynesia recounts the story of a splendid kingdom that was carried to the bottom of the sea. Relying on 10 years of research and extensive travel, Iam offering compelling picture of land of humanity.

## Picture of Submerged Lemuria continent

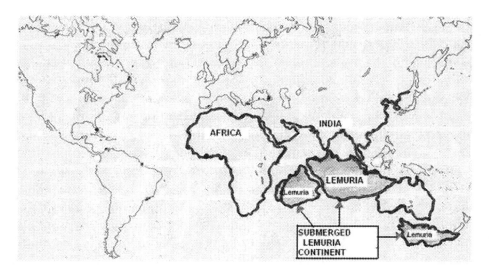

Lemuria continent connecting the Madagascar near Africa and Australia have been such massive land part submerged under water due to Melt down of Polar Ice due to heat waves

When disaster struck Lemuria, the survivors made their way to other parts of the world, incorporating their scientific and mystical skills into the existing cultures of Asia, Polynesia, and the Americas. Totem poles of the Pacific Northwest, architecture in China, the colossal stone statues on Easter Island, and even the perennial philosophies all reveal their kinship to this now-vanished civilization.

The Lemuria theory disappeared completely from conventional scientific consideration after the theories of plate tectonics and continental drift did not happened actually. According to the theory of plate tectonics (the current accepted paradigm in geology), Madagascar and India were indeed once part of the same landmass (thus accounting for geological resemblances).

Some Tamil writers such as <u>Devaneya Pavanar</u> have tried to associate Lemuria with <u>Kumari Kandam</u> (Tamil name for Lemuria Continent), a legendary sunken landmass mentioned in the Tamil literature, claiming that it was the <u>cradle of civilization</u>. With this ancient literary evidence mention it is clear this submerge of Lemuria happened after the epic battle of Ramayana.

The Civilisation and culturural practice found in lemuria.

The main object of the Lemurians was to develop the will and the power of conception. This was the ruling motive in the education of children. Boys were hardened in the most energetic manner. They had to learn to face dangers, to overcome pain, to perform daring deeds. Those who could not bear tortures or face dangers were not considered as useful members of society, but were allowed to perish in the course of their hardships. The endurance of heat up to scorching point, and the piercing of the body with sharp points, were quite common occurrences. For instance, girls were exposed to a storm that they might feel its terrible beauty with calmness; they had to witness fights between men fearlessly, feeling only admiration for the display of courage, strength and power.

<<<<<<< §µΨΦ∞ЖλπЏδ >>>>>>>

Very similar Continent disappearance under ocean in a different time period is the ATLANTIS CONTINENT

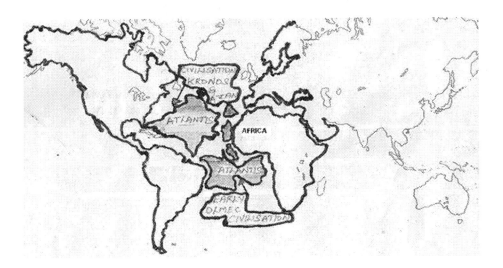

The Archeological evidences confirms the inhabitants of Aryans proved as we saw the habitants from India along with lemurian civilization settlements found till Cambodia and thailand now in the west it is found till atlantis continent one the rich fertile land connecting South America and Africa in atlantic ocean. The wisdom possessed by these leaders, and the powers of which they were masters, could not be obtained through any earthly education, They were worshipped as "Divine

Messengers," and their precepts, commands, and instructions were accepted. By such beings mankind was instructed in sciences, arts, and the construction of tools resulted incomprehensible to the popularity, and did the latter understood the purpose of their great leaders.

To the significance of the places of worship previously mentioned, it was not exactly religion that was fostered there. It was "Divine wisdom and Art." Man felt what was given to him there to be a direct gift from the spiritual world-powers and man perceived and employed natural forces, woman became the first interpreter, even though it differed from the magic of the will on the part of man. Woman was, in her soul, responsive to another kind of spiritual power such as appealed more to the element of feeling and less to the spiritual element to which man was subjected.

The obvious point of perception now gaining clarity which we are seeking in the last chapter the ancestors of Mayan civilization seems to be lived in atlantis continent and the habitants moved to higher range of land like mountain range of Aztec civilization in Mexico and Andean Civilisation pointing North Central costal Mountain Ranges of Peru, Further to the confirmation point, only 2 civilisation in the world confirms the usage and witness of space ships or vimanas, mayan inhabitants further confirming they were the settlers from the southern india and lemuria continents (now called srilankans). The usage of spaceships with Mayan or Andean civilization matching with the geological and archeological proof of land strips found on the high mountain range where the surface has been flattened out for the comfort of landings of the space ship as the NAZCA picture reveals the marvelous ground engineering of air strips like runway path and parking area allotments found,

Further to identify the landing area approach for the high flying space ships, there were identifying marks found nearby and around nazca airstrips, check out these line arts being laid on the massive land area of more than 200 acres of land each, so it will be visible for high flying Vimanas(aircrafts).

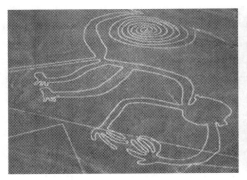

<<<<<<< §µΨΦ∞℥λπ℧δ >>>>>>>

The Ancient Indians making their settlements till south east asian countries and all along the south American continent covering Africa, lemuria, atlantis, Indian subcontinent, North America and the geological and archeological evidences confirming the Brahma temple and Vishnu temple found in grand canyon, its clear Indian habitants settled all over world except the North African region and European region, dominating the migrations continuously happening over thousands and thousand of years. The possession of technology for air travel further confirms the picture below from Mayan paintings and revealing that earlier olmec civilization settlers moved out of submerging atlantis were the forefathers of mayans and andeans.

<<<<<<< §µΨΦ∞℥λπ℧δ >>>>>>>

As the unlocking of secrets revealing Indian habitants taught the world by migrating themselves and educating the uncivilized with the art, culture, governance, trade and disciplined way of life and systemized and structured family system which compelled the attraction of other inhabitants to join the flourishing civilization together producing the strength of the civilisation thus building the Temples and cities all across the settlement area, perfectly matching to the eastern monumental evidence of art and engineering marvel of ANGKOR WAT(Lord Vishnu temple), in

Cambodia. Wherever Indian habitats settled it was the the story of the Tridev's of three supreme deity explained.

Narayana/Vishnu (mythical view—organizer of the creation)/gravitational pull of galaxies in Metaphysics and protons in physics and chemistry and sub-atomic particle of protons in Particle physics.

Brahma (mythical view—creator of universe)/cosmic space and time in metaphysics and neutron in physics and chemistry and sub-atomic particle of neutron in Particle physics.

Rudra (mythical view—destroyer of Universe)/beam of light and heat creating orbital field to keep the planetary sytem in orbital movement in Metaphysics and electron in physics and chemistry and sub-atomic particle of electrons in particle physics.

**Quantum Physics**: **Along with neutrons, protons make up the nucleus, held together by the strong force.** The proton is a baryon and is considered to be composed of two up quarks and one down quark. **As hindus should be very familiar with the following picture depicting above Quantum physics fact.**

<<<<<<< §µΨΦ∞ЖλπЦδ >>>>>>>

Following the Descending saptarishi from Lord Brahma and their lineage till Rama and Lava kusha

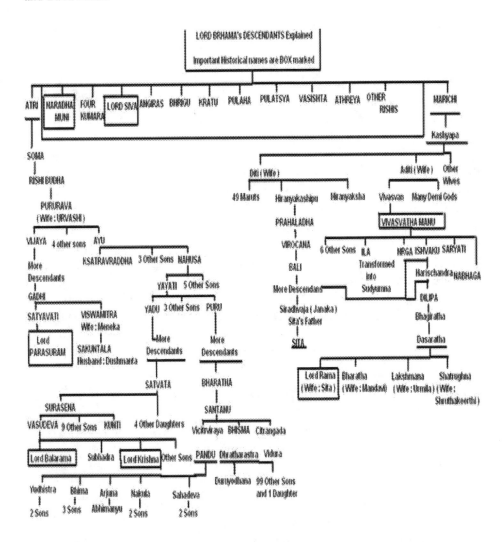

Adi Shankaracharya composed the Bhaja Govindam during his famous pilgrimage to Kashi (Banares). The fourteen disciples are said to have accompanied him. The story goes that when he was walking along the streets of Kashi, he was pained to observe an elderly man trying hard to learn Sanskrit grammar. At his advanced age, the remaining valuable little time of his life should have been used for worshipping the God, instead of wasting on learning a language. This prompted Adi Sankaracharya to burst out this composition, a sort of rebuke to foolish way of living.

bhajagovindam bhajagovindam
govindam bhaja-muudha mate.
sam praapte sannihite kaale
nahi nahi rakshati dukrijnkarane

Worship Govinda, Worship Govinda, Worship Govinda. Oh fool! Rules of Grammar will not save you at the time of your death.

Indian Vedas decoded, still has lot of treasure informations hidden, every hymn in Vedas have inner folded meanings upon understanding the hymns it is the golden pot of nectar where inner folded meanings will keep on unfolding, The revealing facts will be keep on revealing as the most recent proton quarks and proton-proton collision creating such energy discovered in CERN, Geneva in the field of particle physics, was there kept hidden in Vedas was understood by modern science after so many millenniums, but the fact remains that proton quarks is still one of the oldest theory in Indian Vedas.

**Many many incarnations of Lord Narayana is yet to happen in our earth to protect righteousness is believed that will never get counterfeited or defeated as hindus believes and chants as . . . .**

### . . . AUM NAMO NARAYANAYA NAMAHA?! . . . . AUM NAMO NARAYANAYA NAMAHA!! . . . .

### <<<<<<< §µΨΦ∞Жλπ∐δ >>>>>>>

We understood what is meant for Narayana clearly from Vedas and and what it meant for Modern science and **NAMO means i bow or i salute or i respect**? . . . . well **what is AUM**?? Any mantra chanted by any cult any religion that is based from India will chant the mantra or prayers beginning with this word **AUM**?? In Vedas it is explained as PRANAVA MANTRA?!!.

It is first time explained in detail for the benefit of the readers across the world. We will explore the meaning for this word **AUM or pranava mantra** from the view point of modern science as well. To explain this I should at least start it from the womb of the mother.

After the mother is confirmed with pregnancy? do you know how she is confirmed with pregnancy, as the supernova explosion begins?! supernova inside stomach??

Yes the same process of supernova which is happening in space, when the cells gets multiplied and multiplied the life in the mothers womb gets to develop organ by organ, the spinal cord along with brain bulge (head) the spinal cord further grows with nerval grove(Nerves group) as per veda, Till the seventh month the brain is still growing.

Almost at the end of seventh month the formation of brain gets completed, Vedas says till this seventh month the **atman(soul)** has its memory of previous birth, Vedas were the first science to announce the world when the **atman(soul)** is supported with human body **atman(soul)** grows to the size of elbow to the finger tip of the body, once the **atman(soul)** moves out of the body the body becomes dead and the **atman(soul)** upon moving out of the body gets reduced to the size of thumb finger size.

The moment brain growth gets completed during the seventh month (**tip** for the biological scientists to explore or do research further about this detail) there is an invisible gas, forms and sorrounds the fetus in mother's womb?!! This invisible gas is named in Vedas as **SATAKA(M),** which forms during the end of seventh month, which enters into fetus body and makes it to forget, the recordings in the memory of **atman(soul)**, it is considered till this seventh month any **atman(soul)** will remember everything recorded about its previous birth to death and the happening of the after death the travel of the soul for its judgement. The finest example why the Indians relying on Vedas in Hinduism. That is what Vedas and Hinduism is all about the scientific facts and knowledge.

<<<<<<< §µΨΦ∞Жλπצδ >>>>>>>

**That is why I always propose not to differentiante Indian Vedas and Modern science as two different subjects, Indian Vedas is ancient science and Modern science is validation tool for ancient science upon looking into microscopes and telescopes.**

<<<<<< §µΨΦ∞Жλπצδ >>>>>>>

On the birth of the child you can observe closely the breathing style of the baby, is slightly different from adults. An Adult while breathing (note carefully), (one more tip to biologists and doctors from Indian Vedas) while inhaling thru nose the stomach expands out and while breathing out the stomach shrinks in, which is exactly opposite in the cases of babies upto 12-16 months from the birth of the babies?!!???

Watch this breathing movement while infants were sleeping to confirm yourself?!!. It is applicable to all the living beings in humans, animals and underwater mammal species in this world. You know what is the first sound uttered by all animals, human do???? its **AUM?!! AUM?!!.** Westerners say it is mama, as they think the infants were calling their mother as mama?! Sorry Nope?! its **AUMa?!!** not mama. Many misunderstood infants sound uttering **AUM** as **mmm or amma or mama**.

This first noise happens to be AUM(a) due to the inverse breathing the infants do and it is the practice because the inverse breathing happens when the infant inside the womb of the mother. Just think this the kind of scientific informations what ancient Indian Vedas had given us?!?? there were many more scientific or (**biological— bio logical**) engineers lived in ancient india long back and explained Indians to understand what the Vedas kept in it.

Do you know the other explanation Vedas gave about infancy is what this inverse breathing does? (Perhaps the modern doctors could explain this theory better to the critics and aeithists much more clearer) it make the kidney to pump **more growth and resistance hormones**, which makes the growth more faster? for example, Is the growth found in children between age from the day of birth to 10 years is faster?? or after 10 years to 20 years is faster?? which period is faster?? This inverse breathing concept and the timing of the inverse breathing techniques is what **pranayama in YOGA teaches**.

Indian YOGA is the system of exercises which explains about the (**the nitra, meditational practices for concentration basically for MIND, Aasanas the Exercise Postures for the muscles and Nerves system for BODY and Inverse breathing for SOUL) it is the integrated ancient training system for MIND, BODY and SOUL for self realisation.** A technically correct, regular and disciplined YOGA practicioner will feel that the necessary energy or resistance power generating and feeding the MIND, BODY and SOUL is within itself.

<<<<<<< §µΨΦ∞Ж入πЦδ >>>>>>>

*There are lots and lots of matters of science from Vedas were yet to be understood and discovered? As a proud indian i welcome the scientists and researchers from all profession and discipline to take lead from Indian Vedas will sure support as root providing theoretical science to be explored and materialized . . . .*

# 10

# *The theory of world*

However it happened so far, with all true forces of nature, it is clear that man could never outfight the nature or even to the matter of substance of nature proved all along history,

Often humans misunderstood the concept of earth and its resources to the deepest foolishness, and assumed the earth is his potential for his life survival and thinks to himself that everything is granted that he can consume everything he needs. **Humans should understand something really well, that this ecological system of human living along with the balances of nature should never ever be disturbed severly for the matter of consumption, otherwise nature will end up in realigning and rebalancing all its properties, according to its comfort, strength and force this being universal code of law is being unchanged for the billions of years as we know since the life beings started flourishing on this earth.**

<<<<<<< §μΨΦ∞Ж入πЦδ >>>>>>>

**How Hinduism deals with Ecological system?**

Hinduism is a religion always feared and respected the nature. Infact, Hinduism instructs its worshippers to see God in every object in the Universe. Worship of God in air, water, fire, Sun, moon, stars, and earth is always recommended. Indian Vedas conceptualized the very same matter in most of its hymns.

**Rememberance from Bhagavad gita: Krishna says . . . .**

*"I uphold this Universe . . . . all objects . . . . in this Universe rest on me as pearls as tied with the thread displaying as garland".*

Hindus believe that there is soul in all living organisms. The Hindu religion gives great importance to protecting cattles. It nourishes us through its milk and provides manure to grow our food.

<<<<<<< §μΨΦ∞Жλπμδ >>>>>>>

As we have the level of technology today where computers were analyzing the changing patterns of nature continuously from research centres monitoring 24X7 and giving us the signals of possible outcome and its extremities being measured in parameters, we should be taking the cautious steps on understanding this earth surface and underground is filled with water physically and in the air is water filled in the form of gas, h2o and other chemicals in the form of gas in air and in the form of metals in underground. Consumption in this world today is running high even in underground potential supplies and in surface as well as in air well feeding humans plants and animals. It was humans have lost the vital extra-sensory powers as it was not practiced properly due to deviations in materialistic world voiding all the potential communicative relationship between nature and human, it is not in the case of animals,

<<<<<<< §μΨΦ∞Жλπμδ >>>>>>>

**Can Animals Sense Natural Disasters?**

Animals have keen senses that help them avoid predators or locate prey. It is thought that these senses might also help them detect pending disasters. Several countries have conducted research on the detection of earthquakes by animals. There are two theories as to how animals may be able to detect earthquakes. One theory is that animals sense the earth's vibrations. Another is that they can detect changes in the air or gases released by the earth. There has been no conclusive evidence as to how animals may be able to sense earthquakes.

On December 26, 2004, an earthquake along the floor of the Indian Ocean was responsible for a tsunami that claimed the lives of thousands of people in Asia and East Africa. In the midst of all the destruction, wildlife officials at Sri Lanka's Yala National Park have reported no mass animal deaths. Yala National Park is a wildlife reserve populated by hundreds of wild animals including elephants, leopards, and monkeys moved to highlands as they could avoid water flooding in lowlands. Researchers believe that these animals were able to sense the danger long before humans.

<<<<<<< §μΨΦ∞Жλπμδ >>>>>>>

## GLOBAL POPULATION—A MAIN FACTOR

Growing world population and the availability of resource and potential did started to loose balance and need and scarcity in availability of food for the current population is challenging and the pure water resources is compelling to look on nature's urgent call signaled to humans to look and take the matter for consideration. We need trees for rainfall and vegetation, a similar specimen is to be planted before cutting a tree. The vegetation(Plants and Trees) plays a very important role for the balance of nature, also being the major potentials for rain and climatic conditions, More trees meant more supply of oxygen, as trees and plants inhales carbon di oxide and exhales oxygen as nature's distributor of oxygen to humans and humans do the opposite?!! More humans and less vegetation is the slip in balance of ecological environment in simple understandable words the slip in demand and supply proportions. There are lot of germs and worms species helping to the fertility of trees and land as we know, supporting the ecologoical system.

<<<<<<< §µΨΦ∞Жλπμδ >>>>>>>

**Need for Agricultural research and development?**

The available resource of technology and research, scientists were working harder in the areas of bioengineering, genetical engineering and geological fertilization to help agriculture flourish and to increase the production to supply the need of growing population, Like UNO and UNESCO United nations has come out with WFP (World Food Programme) an appreciatable initiative nurturing the need for global consumption and helping the poor nations in supplying food supplies. All funding sources and donors amount accumulated to US $ **3,187,048,425 of 2013** leading nations contributed as per rank **1. America, 2. Canada. 3. United Kingdom, 4. European Commision, 5. Japan**—courtesy WFP. But to fight the imbalance I sencerly believe it is not in contribution to an agency, but each and every nation should put their focus on catering the needs for the development of Agricultural production which also will result in reduction of growing food prices.

<<<<<<< §µΨΦ∞Жλπμδ >>>>>>>

**Using alternate sources of Power?**

Chinese appreciatably took the first initiatives going green by incorporating solar energy potential for consumption for the industrial areas and production despite the nation having the largest population in the world. Supplying with solar electricity to balance the short fall in the power and electricity resources, Many nations took the initiatives for solar power, water and wind power generations for their local usages, japanese have their plans going green in all of their government offices and government activities the efforts initiated is in very large scale in developing the technology after japanese people witnessed the causes of high radiation emission led to the cancer problems from fukoshima nuclear power plant disaster.

<<<<<<< §µΨΦ∞ЖλπЦδ >>>>>>>

**Geopolitical tensions fueling the imbalance of ecology?**

Yes, It is reported that after 1990 due to the raging geo political tensions war between nations causing the movement of seismic plates and causing earthquakes reported more during the later half of the century, the comparison statistical chart from USGS ringing the alarm bells.

Relative annual energy release from earthquakes, magnitude 6 or greater, 1900-2010

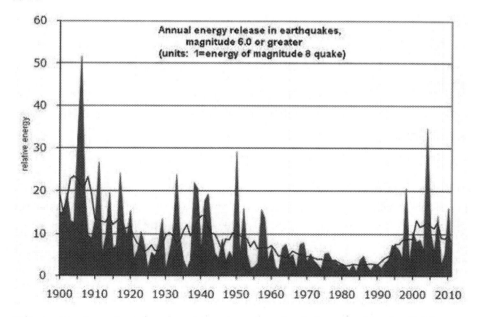

Estimated total annual earthquake energy release (magnitude 8 earthquake = 1 = 1,000 magnitude 6 earthquakes) in red; 7-year average in grey.

<<<<<<< §μΨΦ∞Жλπμδ >>>>>>>

**Extracting oil and other minerals disturbing the seismic plates?**

The chain effect of growing population –> growing automobiles –> growing consumption of crude oil. It is estimated as per reports approximately 72 million barrels as per daily production of crude oil all around world contributing its part of source for generating seismic disturbances, so much crude oil extracted on daily basis causing the seismic plates adjust itself to the gaps from where so much oil extracted causing the tidal disturbance in ocean leading to el-nino and el-nina. Why crude oil so important? As crude oil has so many by products like petrol, diesel, keresone, jet fuel, heating oil, bitumen etc., I agree we cannot surpass consuming oil, but trying to reducing or saving the consumption will help in conservatives. Automobile industry now realized the need for alternate source of energy, trying to design and manufacture the electric and battery cars. Other minerals like coal, lead, bauxite, copper, aluminium were the other minerals turned commodities playing their part in seismic disturbances.

<<<<<<< §μΨΦ∞Жλπμδ >>>>>>>

**Disturbances in Ocean water ecology?**

Ecological consequences of major hydrodynamic disturbances on coral reefs, Reef corals provide the habitat structure that sustains the high biodiversity of tropical reef, and thus provide the foundation for the ecosystem goods and services that are critical to many tropical societies, Integrated predictions from oceanographic models with engineering theory, to predict the dislodgement of benthic reef corals during hydrodynamic disturbances.

Commercial fishing is one of the most important human impacts on the marine benthic environment. One such impact is through disturbance to benthic habitats as fishing gear (trawls and dredges) are dragged across the seafloor. The most disturbed species known to us in common were sea turtles. the Sea Turtle Conservation Project aims to undertake further conservation activities on

- restoring, ecology, and exploitation of sea turtles on the island
- identifying activities that have a deleterious impact on the island's ecosystem

- integrating national and international efforts through advertisement and advocating local community in turtle conservation
- conserving and managing natural resources and habitats that are fundamental for marine turtles
- initiating conservation education programs

<<<<<<< §µΨΦ∞Жλπυδ >>>>>>>

**Why Care About Sea Turtles?**

Sea turtles, are one of the very few animals to eat sea grass. Like normal lawn grass, sea grass needs to be constantly cut short to be healthy and help it grow across the sea floor rather than just getting longer grass blades. Beaches and dune systems do not get very many nutrients during the year, so very little vegetation grows on the dunes and no vegetation grows on the beach itself. This is because sand does not hold nutrients very well. Sea turtles use beaches and the lower dunes to nest and lay their eggs. All the unhatched nests, eggs and trapped hatchlings are very good sources of nutrients for the dune vegetation, even the left over egg shells from hatched eggs provide some nutrients.

—Courtesy Sea turtle conservation and USGS

<<<<<<< §µΨΦ∞Жλπυδ >>>>>>>

**Ocean Water ecology contributing to the child nutrition and health**?

Fish Cod-liver oil is the most reliable and concentrated food sources of four nutrients that are essential to human health: DHA(Docosahexaenoic acid), EPA(Eicosapentaenoic acid), vitamin A, and vitamin D These four nutrients are needed for:

- Healthy skin
- Strong bones and teeth
- Healthy joints
- A healthy cardiovascular system
- A healthy nervous system and prevention of depression and other mood disorders
- A healthy digestive tract

If you're pregnant or you plan on getting pregnant, it's best that you eat foods with DHA on a regular basis in order to support proper development of your baby's nervous system and prevent depression during pregnancy and after your baby delivery. I am convinced that the vast majority of cases of post partum depression can be avoided just by ensuring adequate intake of DHA.

<<<<<<< §µΨΦ∞Ж入π∐δ >>>>>>>

**Protecting the Tropical Rain forest**:

Experts estimates that we are losing plant and insect species each day due to rainforest deforestation. The more rainforest species disappear, so does disappearance for many possible cures for life-threatening diseases. As per reports more than 100 prescription drugs sold worldwide come from Rain forest trees and plants sources. These tropical trees and plants provides greater resources but information tells that less than 5% have been tested by Botanists and Chemists.

Periwinkle, flower based, is one of the world's most powerful anticancer drugs. It has dramatically increased the survival rate for acute childhood leukemia for rainforest plants since its discovery.

Rainforests help stabilize the world's climate by absorbing carbon dioxide from the atmosphere. Rain Forest trees medicinal value of the plants in the rainforests Excess carbon dioxide in the atmosphere is believed to contribute to climate change through global warming. Therefore rainforests have an important in addressing global warming. Rainforests also affect local weather conditions by creating rainfall and moderating temperatures.

The roots of rainforest trees and vegetation help anchor the soil. When trees are cut down there is no longer anything to protect the ground and soils are quickly washed away with rain. The process of washing away of soil is known as erosion. As soil is washed down into rivers it causes problems for fish and people. Fish suffer because water becomes clouded, while people have trouble navigating waterways that are shallower because of the increased amount of dirt in the water. Meanwhile farmers lose topsoil that is important for growing crops.

<<<<<<< §µΨΦ∞Ж入π∐δ >>>>>>>

**Protecting the atmosphere?**

Atmosperic pressure keeps the air pressure in tact, the ozone layer protects us from ultra-violet rays, the Ionosphere layer allows us to bounce radiowaves off it thus giving us radio communication, it prevents water from vaporising and being lost, by allowing clouds to form and rain again, also helps the regulation of the earths standard temperature to its changing seasons.

<<<<<<< §µΨΦ∞Ж入π井δ >>>>>>>

**Accelerated Phasing Out of HCFCs: What is HCFC?**

HCFC is an acroynm for hydrochlorofluorocarbon, which is a compound composed of hydrogen, chlorine, fluorine and carbon atoms?

In September 2007, the parties to the 19th meeting of the Montreal Protocol, an international treaty designed to protect the ozone layer, agreed to speed up the phasing out of HCFC use. As a result, Hong Kong must eliminate 75% of its baseline HCFC consumption by 2010 instead of the original 65% and reduce 90% of the baseline consumption by 2015. In 2020, a total ban on the import of HCFCs is to be in place while 0.5% of HCFC consumption may be allowed for servicing of the then existing refrigeration and air conditioning equipment in the period 2020-2030.

How as a normal and responsible human being can do about HCFC?

- Buying Air-conditioners that do not use HCFCs or CFCs as refrigerants;
- regularly inspecting and maintaining your air-conditioners and refrigeration appliances to minimise refrigerant leaks.
- recovering and recycling HCFCs and CFCs in air-conditioners and refrigeration appliances when they are serviced; replacing and retrofitting such equipment to operate on non-HCFC and non-CFC refrigerant should also be considered

<<<<<<< §µΨΦ∞Ж入π井δ >>>>>>>

**Controlling CO2 emissions?**

The most well known is the burning of fossil fuels. Carbon that was once stored in plants over millions of years now released in the form of $CO_2$. Most of our energy is produced by burning coal, natural gas or oil. Driving a car, heating your house, using your computer, etc. all causes CO2 Emissions

The warming might melt the icecaps, disturbing the oceanic currents that make Europe and the American west coast have such nice climates. If this current stops, a lot of places become very cold. On the other hand, the permafrost in Siberia could melt, releasing lots of methane that strengthens the global warming process reflecting to what we dealt in our 8th chapter about dark years and era of discontinuance.

<<<<<<< §μΨΦ∞ЖλπЦδ >>>>>>>

**Waste Recycling**?

Recycling keeps waste materials out of stagnation and contaminate air and water and generate poisonous gases. But recycling is an effective way to manage wastes and techniques used in recycling system reduces consuming natural resources and creates green evolution. Using recycled material to reproduce paper, plastic, glass and other products saves energy. Transporting recycled materials uses less energy than extracting and processing raw materials. Using recycled materials also create jobs in green economy.

Toxic chemicals are in air, water and soil, and in our bodies. Ecology has well established and effective programs to clean up and manage toxic chemicals. Major industries today use pulp and paper, aluminum smelting, and oil refining have the potential to be major polluters of the environment. The effective, cheapest and smartest approach to reduce toxic threats is in prevention of usage of toxic chemicals to keep them away from our water systems.

GOING GREEN and STABILIZING THE ECO SYSTEM IS THE ONLY WAY OF BALANCING AND LIVING IN THIS WORLD. WE ARE LIVING IN A SYSTEM WHICH IS INTO MATRIX SYSTEMS. CIRCLES INSIDE CIRCLES INTER JUCTING EACH OTHER.

<<<<<<< §μΨΦ∞ЖλπЦδ >>>>>>>

It is each and every single human's responsibility and it starts from you, in preserving the ecology and environment for clean and green eco for our future generation as animals, birds, even worms, bacteria, fungus and virus were doing their part of contribution in balancing the ecology, the lazy and casual attitude in humans atleast should not practiced. It starts from your home, to your street, city, state and nation, Government is planning to invest in solar panels rechargeable batteries to provide street lights which is cutting cost in energy supply grid system development and reducing the consumption of atomic power generation will protect us from radiation upon failure, Fukoshima disaster will remain in japanese minds and made government to reconsider the establishing new nuclear power generator.

Please do remember its not somebody else's world to take care of ecology and environment, it is indeed your world and you are the part of this ecology and you should have the responsibility of preserving it, Followed by Japan, Hong Kong, Malaysia and Singapore, there are households from other asian countries are setting up the solar panels to get their alternative power supply . . . .

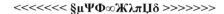

**ENJOY YOUR ECOLOGICAL SYSTEM RESPONSIBLY AND RESPECT THE ENVIRONMENT.**

# OPINION AND SUGGESTION MATTERS

The purpose of this book is not the sharing of my idea, but the purpose is to make someone to understand reaching to the knowledge in depth along with the questions which will enlighten you to obtain knowledge for a purpose or an objective in your life. "You decide your path towards destiny and don't let the destiny to decide your path". Don't wait for the oppurtunities to knock your door, try and create oppurtunities for you to lead your life, It is the way of life you want to lead, Its all about the Journey of thoughts in your mind is going to decide your destiny. everybody has a reason and what is your reason? . . . .

Why do people now reaching to social networking websites and website and mobile Apps advertising today as "STAY CONNECTED", and share your thoughts it is the bigger pool of thoughts shared among in micro seconds across the world and people were willing to stay connected to have their identity, People looking for the identification in the society and expecting their identity to be validated or accepted properly for their own individual objectives???

<<<<<<< §µΨΦ∞Жλπµδ >>>>>>>

**Author is already working on next book on International Finance, the most author deals in his second book is the economic condition of developed economies and emerging economies, its success and problems, Central Bank's Role, Financial status of the world in the notion of Investments and Trading, the logics and ideas behind trading, what is futures and options trading concepts and its controversies, and asset purchasing and much much more . . . .**

<<<<<<< §µΨΦ∞Жλπµδ >>>>>>>

Opinion and Suggestion always differ with each individuals again according to their knowledge and thought process. However author rewards the readers on a simple Quiz with a questionnaire of 10 questions and the first 3 entries of all correct answers every month will be announced as winners with their photograph the procedures of competing on Quiz and the date of announcement of winners and the details of prize money, in author's website (www.authorrajeshji.com)

I welcome all the readers to visit my blogs and my website to post your comments about this book which, as your opinion from readers matters a lot to me,

<<<<<<< §µΨΦ∞Жλπψδ >>>>>>>

when it comes to suggestion, what are all the top 5 topics, most of the authors don't write in Non-fiction and why do you think so?

<<<<<<< §µΨΦ∞Жλπψδ >>>>>>>

what are all the top 5 topics that readers obviously expecting from Non-fiction Authors to start writing? You can also post these suggestions in my blog or in my website.

<<<<<<< §µΨΦ∞Жλπψδ >>>>>>>

Iam looking forward for the authors from various nations who feel this book is good enough to be translated, please feel free to approach author thru email, or website, as author holding © copyright for translation and adaptation rights, however author will provide translation rights to the translation authors who did the translation works earlier known for their reputation and name of the book(s) and publisher details and published year is required to submit for evaluation for considerations for issuance of translation grant . . . . for all international and Indian languages translaters

Please do remind the translaters deviate the guidelines, terms, conditions and rules issued under US copyright regulation may face severe compensation damages and shall undergo all the legal actions and implications that the author proposes and author reserves his rights in the discrepancies and fraudulent if found in reproducing this content.

<<<<<<< §µΨΦ∞Жλπψδ >>>>>>>

*Coming Soon . . .*

*Coming Soon . . .*

## The Journey Of Mind

# Connecting to the Spirit,
# within.......

**RAJESH JAYARAM**